ANSWERS FROM SCIENCE

ANSWERS FROM SCIENCE

Consciousness and the Origin of Life Explained

J.C. MAGHINARD

Website : answersfromscience.com

Contact : maghinard@gmail.com

© J.C. Maghinard, 2022
ISBN 9798455468490
Edition 1.1

Cover photo :
ESA/Hubble

The big answers are in the great laws.

TABLE OF CONTENTS

1. FOR A SCIENCE WITHOUT MATERIALISM · 9

2. HOW SCIENCE WORKS · 13
2.1 The scientific method · 18

3. LET'S BE LOGICAL · 23
3.1 Logical consistency · 26
3.2 Logical impasse and false mysteries · 27
3.3 Artificial division and artificial confusion · 31

4. THE MATERIALIST BELIEVERS · 37

5. THE UNIVERSALITY OF THE LAWS OF NATURE · 43
5.1 Conceptual unification · 46

6. THE IMPORTANCE OF THE INVISIBLE · 55
6.1 The invisible side of nature · 57
6.2 The law of selection · 60
6.3 The invisible and the universality of the laws of nature · 67
6.4 In summary · 77

7. THE IMPORTANCE OF ENERGY · 81
7.1 Under pressure · 87
7.2 Bosons and fermions · 91
7.3 Energy and the universality of the laws of nature · 96
7.4 In summary · 99

8. CONSCIOUSNESS DOES NOT COME FROM THE BRAIN · 103
8.1 Most of the human being is invisible · 106
8.2 The brain is an intermediary · 110
8.3 The law of inertia · 122
8.4 Consciousness comes from energy · 125
8.5 Conscious experiences made outside of the body · 129
8.6 In summary · 133

9. LIFE DOES NOT COME FROM MATTER · 137

9.1 The origin of life according to materialists · 138
9.2 The law of equilibrium · 142
9.3 Most of life is invisible · 146
9.4 The law of reproduction · 152
9.5 The transmission of life on earth · 162
9.6 The origin of life · 166
9.7 In summary · 178

10. THE UNIVERSAL ORGANISM · 181

11. SCIENCE AND REVELATION · 193

11.1 Intellect and intuition · 202
11.2 In the Light of Truth – The Grail Message · 204

12. WHAT IS A GOOD THEORY? · 207

13. SYNTHESIS · 219

13.1 The three pillars of universalism · 220
13.2 The main laws of universalism · 221
13.3 The main unifications of universalism · 225
13.4 The main symbol of universalism · 226

14. THE MEANING OF LIFE · 227

14.1 In conclusion · 237

FINAL SUMMARY · 239

BIBLIOGRAPHY · 241

Physics and chemistry · 241
Biology and origin of life · 244
Neuroscience and consciousness · 249
Science and philosophy · 253

1. FOR A SCIENCE WITHOUT MATERIALISM

The answers to the great mysteries of science are in the laws of nature.

But to see those answers, we must first leave materialism behind.

Consciousness and the origin of life are viewed as the two greatest mysteries of science. Is it not strange that science, the mission of which is to understand how nature works, is confused about such vital matters? Despite incredible progress in many areas of science, how is it that these enigmas have not yet been resolved? Given their great importance, should not these subjects be those that are understood most clearly?

How strange…

The confusion surrounding these questions is so great that many even say that science cannot answer them. Many then advise that we turn toward religion for answers. But as soon as we enter this domain, we find ourselves lost in a labyrinth of contradictory opinions and supernatural beliefs, with the result that our desire to gain natural and logical answers is never satisfied.

So, how do we escape this impasse?

The purpose of this book is to offer answers to these big questions by explaining why scientists encounter so much difficulty when trying to solve the enigmas of consciousness and life, and by explaining how it is possible to solve these mysteries with logical solutions based on the laws of nature.

This goal may seem very pretentious. However, we will see that the solutions are actually very simple and accessible to all. The most important thing is first to change the way we approach these problems because this confusion is only created by false beliefs that we maintain, which prevent us from seeing the answers we already have before our eyes.

There are all kinds of false beliefs, but, let us say it bluntly, the most important misconceptions that prevent science from making progress on these questions come from materialism. Materialism is a worldview that considers consciousness only as a product of the brain, and life only as a product of matter—beliefs that, as everyone knows, currently dominate in the scientific community. Materialists like to portray this attitude as the only scientific way of examining consciousness and life. However, as we will see, it is possible, within the framework of science, to approach these subjects in a completely different way. Not only is it possible, but it allows us to resolve many questions currently considered great mysteries. Science is neutral, it does not force us to be materialists. On the contrary, science works much better when we leave our beliefs aside—that includes materialistic beliefs as well as religious ones.

In this book, we will rebuke materialism. Inevitably, many will think that this attitude comes from a bias against materialism, whereas we will only look at materialism with *the same* critical eye with which we must look at any belief system. This, to show the flaws of materialism, flaws that must convince us that it does not deserve the special treatment that it gets from too many scientists, who neglect to be as skeptical about materialism as they are about other philosophies.

We will not only find fault with materialism, but we will also see an alternative approach, named *universalism*. This approach is so named because it is based on the universality of the laws of nature, a principle that is the central pillar of science, as we will see throughout this book. It is possible to solve the enigmas of consciousness and life in a way that fits perfectly with the laws discovered by science, but to see these solutions we must first abandon the beliefs that consciousness depends on the brain and that life depends on matter.

The idea that consciousness and life do not come from matter is not new, it has been with humanity since the dawn of time in a wide variety of forms. However, the way we will use this idea will be very different from what we usually see. The universalist theory presented in this book is not associated with any religion or spiritual philosophy, nor with any other school of thought that uses this term. Universalism is a set of logical and natural solutions, which answers the big questions by relying solely on the most well-tested laws of science.

To fully understand the solutions that the universalist approach contains, we will first see a summary of how science works, and of its key findings. Then we will go deeper into the subjects of consciousness and the origin of life to understand exactly why universalism can succeed where materialism has failed.

The confusion that exists around consciousness and the origin of life has no reason to be, since it is possible to explain them with the knowledge that science currently has. There is no need to wait for future great discoveries because the answers are not at the forefront of science, but in the *basics* of science: the laws of nature.

2. HOW SCIENCE WORKS

At its core, the practice of science is child's play.

To understand how we can solve the enigmas of consciousness and life in a way that is consistent with science, it is necessary first to have a good understanding of how science works.

At its core, the practice of science is very simple, because its basic rules are very simple. The practice of science can be seen as a game, the goal of which is to correctly combine two types of concepts that are the *elements* and their *logical relationships*.

An analogy that we can use to describe this process is a construction set made of balls and sticks, similar to those that chemists use to represent molecules. In this analogy, the elements are the balls, and the relationships are the sticks. The goal of science is to combine these pieces in a way that corresponds to reality, and it is the experiments that tell us if we have the right answer or not.

The instrument we use to do this work is the intellect, our ability to reason. Science comes from the activity of the intellect, the job of which is to form representations of reality by manipulating these two types of concepts that are the elements and their logical relationships. The fruits of this activity are like maps, plans or formulas, and we use these mental patterns to explain the world and make predictions.

Each of us uses representations of this type that our intellect has elaborated over time, representations to which we give several names, such as "hypothesis," "theory," "concept," "model," "idea," "philosophy," "worldview" or "belief system." Furthermore, just like a map can be good or bad, these representations can be true to reality or filled with errors if the work has not been done correctly.

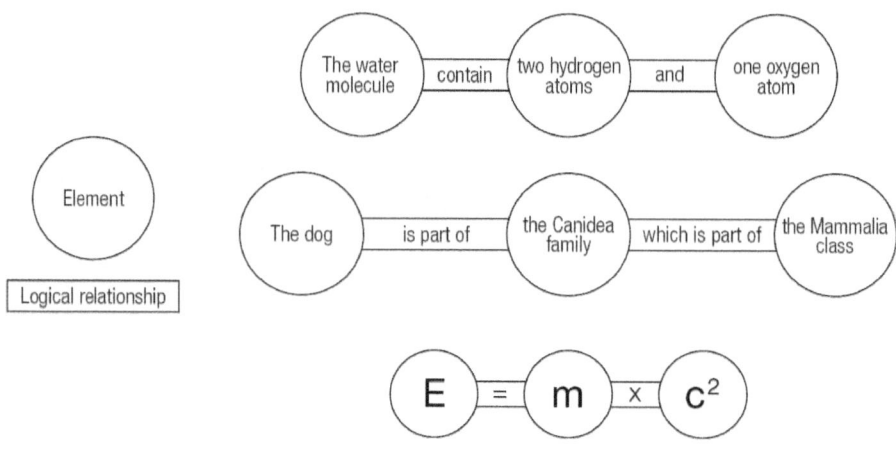

Examples of mental representations borrowed from chemistry, biology and physics. These can be seen as "molecules of knowledge."

The goal of science is to build good representations of reality. We can achieve spectacular results with this practice, but that does not change the fact that, at its core, it is a simple process used by everyone. In any field of activity, when you specialize and put in a great deal of effort, you can achieve extraordinary results; just as a professional athlete can surprise us with his prowess, by practicing a game accessible to children. It is the same with the practice of science. The results are often spectacular, as demonstrated by the many technical achievements made possible by science, but that does not change the basic simplicity of the process: *It is only a matter of determining what the relationships between the elements are.*

When we discover a constant relationship between categories of phenomena, this relationship can be formulated as a law of nature. Because these laws encompass many elements, they can be used to explain a great deal. The laws of nature are the most important discoveries of science, as we will see throughout this book.

What is nice with logical relationships and laws is that they are *always simple.* This statement may seem surprising, because we all have seen images of scientists who fill their blackboards with seemingly incomprehensible formulas. However, when we look more closely, we realize that their formulas contain only simple logical relationships, which are, in fact, the same ones we use to reason in everyday life. Scientific geniuses construct their theories with the same basic logic we all know, just as a virtuoso musician uses the same notes as a beginner. In fact, scientific theories become complex only because they contain many *repetitions* of these simple relationships.

While admiring a cathedral, we can be in awe looking at such a technical feat, and this can make us forget that building such a structure is essentially stacking blocks, and repeating this gesture many times, another activity that children do too. It is the same with the great scientific theories. They are the result of centuries of work by researchers who have managed to combine a large number of elements, just as we combine stones to build a cathedral. The great theories of science are remarkable achievements. They may even be considered like the cathedrals of scientific thought; but the construction of a theoretical edifice consists only of connecting elements with the right logical links and the right laws, repeating this simple operation many times, just as one puts blocks on one another.

Intellect and science are interested in logical relations, but we must first know the elements between which to establish these links. These elements are provided by another area of knowledge, *intuitive knowledge,* which can also be called knowledge by experience. In a scientific context, we also speak of empirical knowledge. The intellect alone cannot produce any science, because it can only analyze

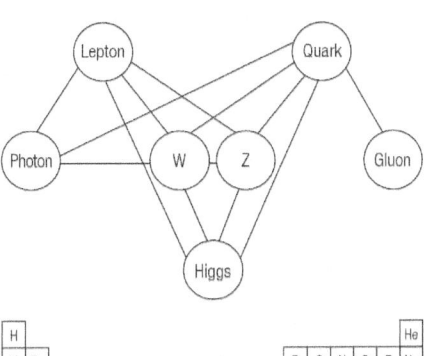

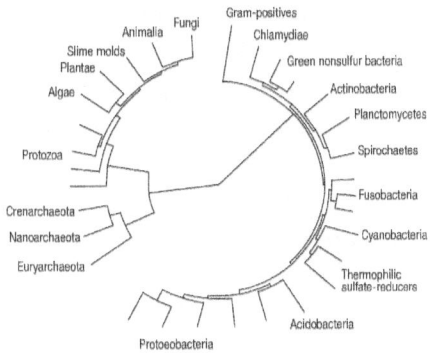

Figures from major scientific theories. From top to bottom: the interactions within the standard model of particle physics, the periodic table of the elements and an evolutionary tree. Regardless of the complexity of a theory, the principle remains the same: the goal is to establish the logical relations between a wide variety of elements.

the links between the elements, it is not able to experience them. The intellect can classify phenomena only *after* we have gained the intuitive knowledge of them through the experience of reality.

To illustrate this, we just need to observe a child being raised. If you want to teach a child what a cat is, you do not read the definition of the word "cat" from the dictionary, because the intelligent way to proceed is not to go through the intellect. Quite simply, you only need to show this child a cat and say that it is a cat, and the child will understand instantly what this animal is better than you could have done using all of the words in the dictionary.

This shows that all the concepts and words used by the intellect are worthless if we do not *first* have the experiences to which they refer. No one can really understand what the word "cat" means without having seen one before! Just as no one can really understand what the word "red" means without having seen this color, that no one can know the taste of a meal without having eaten anything like it, and so on… In this sense, words only serve to label experiences, they can never replace them. It is the same with numbers, which are also a limited way to approach reality. Trying to understand nature with mathematics as our only tool is like trying to understand a creature by studying only its skeleton!

Although they are often underestimated, the limits of intellectual knowledge are known to all. For example, if someone wants to learn how to drive, we do not just ask this person to read a bunch of books on the subject before sending him or her alone on the road, considering that this person now "knows" how to drive! Because we know that the only way to learn to drive is to put your hands on the wheel to gradually build up driving *experience*, this with the help of an experienced driver. It is the same with learning a trade. Theoretical training is never enough, you always also need to gain practical experience.

You can spend your life studying in books what an apple is, you will never know it as well as by eating one! Thus, intuitive knowledge shows its distinct nature compared to intellectual knowledge. Only intuition can truly know reality, by directly experiencing it, and these experiences are far too rich to be wholly grasped by concepts, numbers and words. With the direct experience of reality, we instantly gain a phenomenal amount of information, and we do so with a precision that is impossible for the intellect to reach.

We can even go further by saying that only lived experience is true knowledge, whereas what the intellect calls "knowledge" is only a set of more or less precise representations of reality. These representations are like portraits and have the same limits. A portrait, even a very good one, is never the equivalent of its subject. In the same way, scientific theories are nothing much compared to the true richness of the nature that they try to describe.

The role of intellect and science is not to understand the profound nature of the elements they study, but to grasp the logical relations and the laws that exist *between* these elements. With this understanding, the intellect forms theories and models that allow us to understand better the functioning of the world. Intellectual understanding is useful because once the links are well understood, they allow us to make deductions, predictions, plans...and all this helps us to have some control over reality.

Thus, the activity of the intellect balances the activity of intuition, because intuition exists in the present moment and is incapable of planning anything. Without the intellect, intuition would be unable to function in everyday life. In this sense, the intellect can be seen as a *navigation system*, and the models it creates tell us the means we need to use to achieve our goals, just as roadmaps do.

If a theory accumulates errors, it can only lead us into a dead end, because it is a wrong model of reality, like a poor map. As we will see throughout this book, this is precisely what materialism is: a false representation of reality that leads us directly into an impasse.

Materialism is like a poor map. It can only lead us into dead ends.

2.1 THE SCIENTIFIC METHOD

We have seen previously that the purpose of scientific activity is to understand the logical relations that connect phenomena. Therefore, the scientific method is only the means we use to find these relations or these laws. It is not a practice that can be engaged in by experts alone because this method only reproduces the natural functioning of the intellect. Indeed, as soon as a question arises about a phenomenon, the intellect cannot answer it by any other means than the "scientific method."

The scientific method can be summarized in four steps:

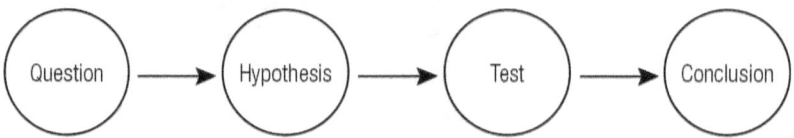

We automatically follow these steps when we reason, since this is the path that our intellect naturally follows when it needs to answer a question. To see how we all use the scientific method, let us use the example of an ordinary problem: a leaky roof.

Our scientific investigation begins with a *question*: "What is the cause of this leak?" The question determines the type of logical relationship to be established. In this case, it is a cause and effect relationship that must be explained between the effect we observe, the drops falling from the ceiling, and a cause still unknown.

The second step is the *hypothesis*. At this stage, the intellect draws from the knowledge it possesses, trying to determine what the most likely causes are. For example, if we know that there is no plumbing near this area, we will be sure that the water comes from outside. If the phenomenon has never happened before, we will consider that it is because of a new factor. If there is a rainstorm with a great deal of wind, it means that the water infiltration may have something to do with this storm, and so on.

Reflecting like this, the intellect seeks the most likely answers. In the case of a leaking roof, the reflection preceding the hypothesis may last only a few seconds. In the case of more complex scientific problems, such as diseases with poorly understood causes, this can take years. Because, in more complex cases, our already acquired knowledge is insufficient to form a potentially correct hypothesis. The construction of our hypothesis must then be preceded by a period of research that can be very long.

The most likely assumption will usually be considered first to proceed to the next step, the *test*. The test is an observation of reality, which intends to verify the predictions that our hypothesis allows us to make. For example, if we have established that the most likely cause is that there is a hole in our roofing because the wind has torn away shingles, to validate our hypothesis, we need to go on the roof to check if there is really a hole. So, we take our ladder and climb onto the roof, braving the storm, putting our health at risk in the name of science.

Once on the roof, if we see that there is indeed a hole, that confirms our hypothesis, and we are ready to draw a conclusion…or maybe not. Indeed, the observation of this hole might be a satisfactory answer in everyday life, but, for a scientist, things are not so easy. For the observed element to be confirmed as the cause, it is also necessary that *all other possibilities* have been excluded. This is the main reason science needs laboratories, to control the experiment to eliminate all the possibilities except those we want to test. This is what scientists call "controlling the variables." Regarding the water infiltration problem, a picky scientist would consider that it is also possible, although unlikely, that the rainwater also enters through another place. To really confirm the hypothesis, it would be necessary to repair the roofing, wait for a similar storm to be repeated, and see if water infiltrates again. If it does not infiltrate anymore, we can move on to the next step.

The *conclusion* stage consists simply of reporting whether our hypothesis has been validated or refuted by our experiment, the test of reality, and to establish the logical relationships involved. In this case, we would have proved that there is a cause and effect relationship between a hole in the roofing and water seeping into the house.

Obviously, this is not a discovery that can win one a Nobel Prize, but we must not forget that all of science's conclusions, once clearly established, are never more complicated than that! The link between a hole and a water infiltration is simple, as is the link between energy, mass and the speed of light, which can be expressed by a simple mathematical formula, $E=mc^2$, a fundamental equation, which is nevertheless used continually to symbolize scientific genius. *All logical relations are simple*, and it is not because these links are obscure to us that they are complicated. It only means that we do not have a habit of thinking about them. You can form theories containing thousands of logical links of all kinds, if you consider them one by one, they remain as easy to understand as "2+2=4."

Of course, a leaking roof is an easy problem to solve when compared to the sometimes-difficult problems that scientists face. If you want to validate a hypothesis predicting the presence of a hole in a roof, a person equipped with a ladder is enough; if you want to validate a hypothesis predicting the efficacy of a new drug, you need years of clinical trials; if you want to validate a hypothesis predicting the existence of a new particle, you need thousands of scientists, millions of dollars and a particle accelerator… The difficulties in science arise when the hypothesis is difficult to test for technical reasons or because there are a large number of possibilities to sort out. Arriving at a definitive conclusion can then require decades of work.

We use the same method to solve everyday problems that professional scientists use to solve the most important issues, the main difference being that we apply it less rigorously. This process has been used since ancient times by hunters, gatherers and farmers of the distant past, who had to understand their environment to survive.

The sequence "question, hypothesis, test, conclusion" is only a summary of the scientific method. There are many ways to present it, but the principle always remains the same: The intellect forms *representations* of reality, which it seeks to improve continually using the information provided by our *experience* of reality. This in the same way that an artist gradually draws a portrait by alternately looking at his work and his subject. It is only through this constant interaction with reality that our good mental models can become ever more precise, and that the bad ones can be corrected or rejected completely.

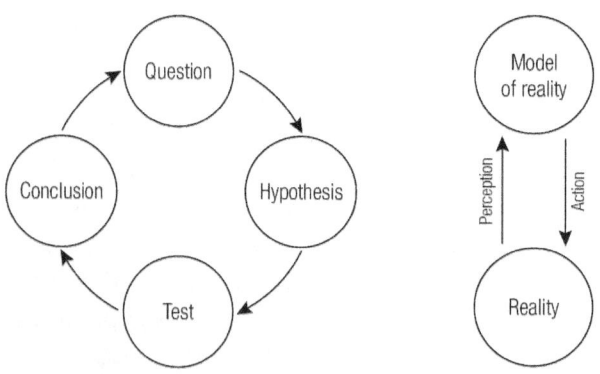

The scientific method can be seen as a loop. It is a process that our intellect continually repeats to check the accuracy of our mental representations.

Thus, the scientific method can be seen as a *loop*, a circular succession of stages that our intellect continually repeats to ensure that our representations are always consistent with reality. The conclusion of a test often brings new questions and, by repeating the scientific method many times, we gradually build an ever more precise theory on the subject under study.

In summary, we can consider the scientific method as a control process, that the intellect consistently uses to check the accuracy of our mental representations. This process allows us to discover and eliminate the errors in our models, errors that can be of quite varied origin: they may be false beliefs of all kinds, personal interests that make us prefer one theory over another, miscalculations, poorly calibrated devices, poorly designed experiments, poorly controlled variables, ignorance of certain facts, misinterpreted observations and many other things…

This hunt for errors is the reason all the steps leading to a scientific discovery must be scrutinized and repeated by several groups of independent researchers. False premises are the great enemies of science, enemies with whom it fights endlessly because they are everywhere and spread like weeds. Even the circles formed by the scientific elite are not safe from this evil.

The scientific method requires us to be continually vigilant, especially regarding our personal beliefs, to avoid relying on false models of reality. For when we begin to value our beliefs more than facts, our intellect can no longer correct our mental models. In this state, which is a mixture of pretension and intellectual laziness, we value only what seems to confirm what we already believe. This is how our misconceptions disconnect us from reality because every false belief we adhere to adds another brick to a wall that we build between ourselves and reality.

To deconstruct this wall, we must be ready to abandon our prejudices, to listen to what the discoveries of science really tell us. Learning to examine our beliefs does not mean that we must live in perpetual doubt, but rather, that we must not hesitate to question ourselves, to build our vision of the world only on facts and laws that are rock solid. We must test our most important beliefs in all possible ways until they give in or become convictions. As we progress this way, we realize that very few of our beliefs resist these trials, but what remains is entirely trustworthy. We can then build on this base a vision of the world rooted in reality.

Every false belief that we adhere to adds another brick to a wall that we build between ourselves and reality.

3. LET'S BE LOGICAL

The root of all problems is that we are more concerned with protecting our false beliefs than with seeking logical solutions.

Logic possesses an extraordinary power: the power to make everyone agree on a subject. If you travel the world with the equation "2+2=4" written on a piece of paper, you will not find anyone who disagrees with this. The materialistic intellectual, as much as the religious fanatic, will both tell you that this equation is true. This is the power of logic, it can bring together people who have completely opposing worldviews.

When an answer comes from basic logic, we all have the same opinion about it. So, would it not be wonderful if the answers to the great mysteries of life were also in basic logic? This could put an end to all the conflicts that arise because, on these issues, people have different opinions about what is true or false…

The position upheld in this book is that the answers to the mysteries of life are indeed in basic logic, and those questions should never have been considered like enigmas. These mysteries are illusory. They come from erroneous interpretations that have become false beliefs. As we will see, we find errors of this type *both* in materialistic and religious philosophies—no one is above this evil. Hence, the importance of correctly understanding its origin.

First, let's see a summary of how logic works to understand better how our reasoning can be disrupted by our false beliefs. Logic is the set of rules that the intellect uses to distinguish truth from falsehood. The rules of logic function similarly to those of mathematics since mathematics is only one particular subset of logic. In philosophy, it is even common to break down reasoning in a way comparable to a mathematical formula, for maximum clarity. The elements on which an argument is based are decomposed into premises, which are reasons that serve to support a conclusion. Here is an example of this type of formulation:

Premise 1: If I do not fix my roof, there may be a leak in my house.
Premise 2: I did not fix my roof.
Conclusion: Therefore, there may be a leak in my house.

If we extract the formula of this reasoning, it gives this:

Premise 1: A implies B
Premise 2: A is true
Conclusion: Therefore, B is true

This formula can also be represented with symbols, such as a mathematical equation, or as a diagram:

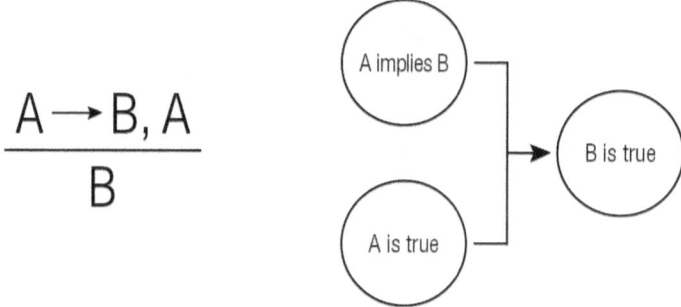

As with the scientific practice we have seen in previous chapters, the basic rules of logic are simple. These rules are called "rules of inference" by philosophers. There are several of them, and they have all kinds of fancy names, like the previous example, which is called "modus ponens." Each of us follows these rules in our reasoning, even without having studied them, since they only reflect how the intellect works. Despite the simplicity of its basic rules, logic is a vast field, and we will only scratch the surface here. The bottom line is that well-constructed reasoning works like a mathematical formula, where the premises and the conclusions follow one another automatically, obeying simple rules.

The conclusions of one reasoning become the premises of other reasonings, and this is how the intellect constructs logical networks that can become complex, such as this example:

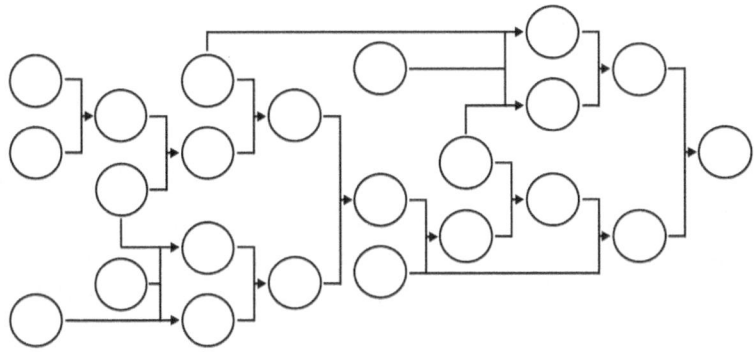

Obviously, it would be very bland and boring if we always presented our arguments in the form of diagrams or formulas, as in mathematics, but we must be aware that any reasoning can be deconstructed in this way, where elements follow one another clearly by obeying strict rules. This is also the way computers work, where all operations are broken down into very fast successions of logical steps. We then realize the importance of ensuring that no false premises are inserted into our reasoning, because the intellect can take a wrong direction by relying on them, to construct representations disconnected from reality.

If we go too far in the wrong direction, we end up believing that the world is filled with impenetrable mysteries, whereas this confusion is only the artificial product of our false beliefs, beliefs that we cling to for all kinds of bad reasons. All of those points will be deepened in this book, especially concerning materialistic philosophy and the most important of its false premises: "consciousness depends on the brain" and "life depends on matter."

Now, let us look at other important notions about how logic works, notions that will be used throughout this book.

3.1 LOGICAL CONSISTENCY

For reasoning to have logical consistency, the same premises must always bring the same conclusions. When similar premises lead to different conclusions, it means that the logic is inconsistent. This is a sign that there is a false premise somewhere, which often comes from a false belief that one wants to keep, and the price to pay to keep our false beliefs is always that of logical consistency.

For example, many religious fundamentalists do not believe in the evolution of species because it contradicts their interpretation of certain religious texts. Alternatively, they accept the idea that an animal species can be transformed when their breeders choose to reproduce only the animals that meet specific criteria.

This logic is inconsistent because the two cases are similar; the only difference is that, in the wild state, it is nature that operates the selection. This is because the animals that best fit their environment are also the ones with the highest reproduction rate, so they naturally tend to become models for later generations. The principle of selection in artificial breeding and that of natural selection is the same, only the form is different; therefore, it is inconsistent to believe that, in nature, species are fixed and to accept that in the other case, they can change.

In short, our reasonings are inconsistent when we treat differently subjects that should logically be treated in the same way. It is this kind of logical fault that is denounced when someone is accused of using a double standard.

Another well-known example of incoherence comes from political corruption, when friends of those in power manage to obtain preferential treatment to which they should not be entitled if the laws were always applied impartially.

False premises and inconsistencies are sometimes referred to as "logical flaws." As a defect in the construction of a building can ultimately cause it to collapse, the presence of a single flaw at the base of an argumentation may be enough to invalidate all the conclusions that were based on this fault.

For example, it would be enough for consciousness to really be independent of the brain, to bring down the logical building of materialism. Because the idea that consciousness is only a product of brain activity is one of the basic premises upon which their worldview is constructed.

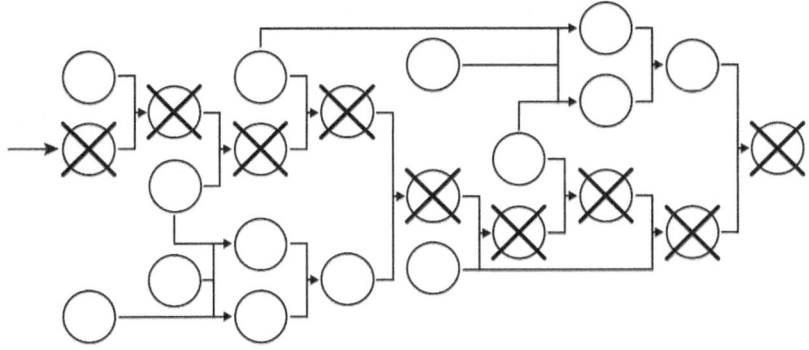

Logical flaws: A single false premise at the start of a logical network is enough to invalidate all the conclusions that are based on it.

3.2 LOGICAL IMPASSE AND FALSE MYSTERIES

When several conclusions produced by our reasoning are irreconcilable, we fall into a logical impasse. In other words, when one believes many things that cannot be true at the same time, it is a situation that our intellect cannot handle properly.

Falling into a logical impasse is inevitable when inconsistencies are inserted into our reasoning. Faced with such an impasse, the only solution is to change or give up the false premises that led us into this dead end. Otherwise, we cannot go further.

When maintained, logical impasses can create false mysteries. We are in the presence of a false mystery when a question can be solved with the knowledge we already have, but we do not see the answer because we *believe* it to be different from the solutions we have at hand. In short, we create a false mystery when we believe that there is a difficult problem outside of us, whereas the origin of this difficulty lies only *within* us in our interpretations and beliefs. In these cases, the "mystery" is only the logical impasse created by our contradictory beliefs; it has no real existence and must be seen as a false mystery or an artificial mystery.

If a true mystery can be solved by the acquisition of new knowledge, an artificial mystery can only disappear by changing our approach to the problem—simply because this type of mystery is created *by* our approach to the problem! This type of enigma often begins with a legitimate question that appears insoluble to us because of a wrong approach, which has the effect of creating a mist of artificial confusion that prevents us from seeing the answers as long as we do not change our approach.

To illustrate this situation, we can observe what happens when we see an illusionist's show. The different techniques used by illusionists are essentially designed to create artificial mysteries by making us believe that what is going on is something other than what is actually happening. Therefore, they are quite appropriate to illustrate what is a false mystery produced by our beliefs.

Take the classic example of the illusionist who disappears in a smoke screen at one end of the stage, only to reappear immediately at the other end in another smoke screen. We are apparently in front of a mystery: "How did he move so fast?" As long as we approach the problem this way, considering that the illusionist really moved like that, we are in a logical impasse, because this interpretation conflicts with our conception of reality, which tells us that it is impossible to move so quickly, and it is this incoherence that creates the sensation of mystery.

We can think about this issue for years and imagine all kinds of fantastic solutions. As long as we continue to believe that the illusionist really moved that fast, we will not find any answer. This so-called mystery will persist as long as we do not change our approach to the problem, renouncing the belief that the illusionist has really instantaneously moved to the other end of the stage, even if that contradicts appearances.

The answer to this false mystery comes easily as soon as we decide to listen to the laws of nature that tell us that it is impossible. Since the illusionist could *not* have moved so fast, it means that the person at the other end of the stage could *not* have been the illusionist, even if it looked like him. So, it could only have been a look-alike, who came out of his hiding place while the real illusionist left the stage, and this sleight of hand was masked by the smoke!

The mystery of the teleportation of the illusionist seemed opaque to us only because we believed that it was really what unfolded before our eyes, and this mystery disappeared by finding an easy answer when we stopped believing the appearances. Above all else, illusionists' tricks rely on mental manipulation intended to make us misinterpret what is happening before our eyes, and it is these false interpretations that then give rise to the sensation of mystery when we believe them, that is to say, when one begins to value them.

This illustrates the fundamental role our interpretations and beliefs play when we attempt to answer a question, by showing that a wrong approach based on false beliefs is enough to make a problem appear insoluble. In these cases, the more the false beliefs behind the wrong approach are maintained with intensity, the more the "mystery" will seem insoluble.

As we shall see, this situation is precisely the one in which materialists find themselves regarding the enigmas of consciousness and the origin of life, questions that they themselves call the "greatest mysteries of science." How does the brain generate consciousness? How could matter have generated life? Materialists do not find answers to these mysteries because this is simply not what happens in reality!

If materialists are *not* able to answer the question, "How does the brain generate consciousness?" it is because consciousness does *not* come from the brain. Moreover, if they are *not* able to answer the question, "How could matter have generated life?" it is because life does *not* have its origin in matter. These false mysteries are created only by materialistic beliefs and disappear as soon as we listen to the laws of nature that tell us clearly that such phenomena are impossible, without worrying about deceptive appearances.

The belief that consciousness and life are dependent on matter has led materialists into logical impasses in which they have been stalled ever since. However, the materialists are only victims of illusions that are easily explicable as soon as we change our approach to the problem, considering consciousness and life as phenomena that do not have their origins in the brain or matter, even if from a certain point of view it looks so.

The trap into which materialists fall is the same that we fall into when we believe that what the illusionist is showing us is really happening. It is a trap as old as the world itself: *the trap of appearances.*

It is the same trap into which the peoples of the past fell, those who believed that the Earth was flat because it looks flat, or that the Sun revolves around the Earth because it is so obvious that you have to be crazy to question it. Nature can produce very convincing illusions, and no one is so knowledgeable to the point of being able to always avoid this trap! The mistake many make is to believe that the instruments of science are so advanced today that we are now assured of no longer falling into the trap of appearances, when in reality there are still many false beliefs based on illusions that are promoted by the intellectual elites of our time, as was always the case in the past. Indeed, is it not pretentious to believe that this problem that has always been with humanity is finally solved today?

Those who believe today that consciousness comes from the brain and

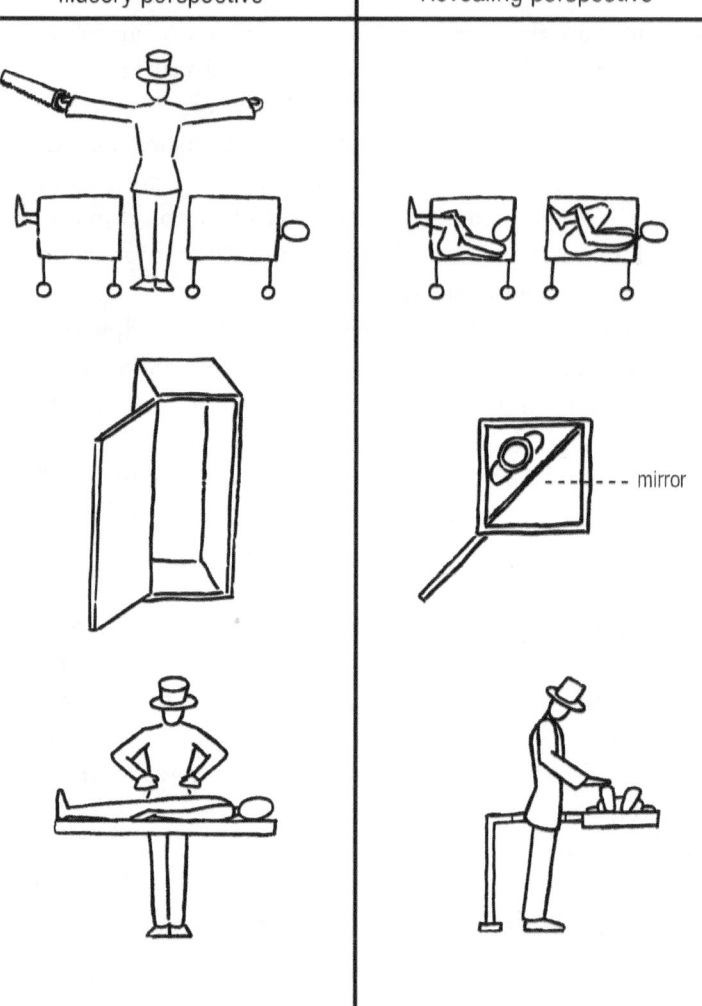

If one looks for an answer to the question, "How did the illusionist cut this person in half?" there is no answer because no one is cut in half. The same applies to the question, "How did he disappear?" because no one disappeared. And the same is true for the question, "How does this person levitate?" because no one levitates.

It is the same for the enigmas of materialism. If one seeks an answer to the question, "How does the brain generate consciousness?" there is no answer, because the brain does not generate consciousness. The same applies to the question, "How does matter generate life?" because life does not have its origin in matter. All these false mysteries come from beliefs in illusory appearances.

that life comes from matter will one day be considered in the same way as those who believed that the Sun revolves around the Earth. Just as illusionists' tricks only work because the viewers have a limited view of the scene, the appearances that impress materialists can do so only because the instruments of science give us a limited vision of reality. However, as we will see, this limitation is not reason enough for adopting materialistic beliefs, because by making logical deductions based on natural laws, we can understand very well what lies beyond the limitations of the instruments of science. It is not a continual improvement of the instruments of science that can free us from the trap of appearances, but an ever-deeper understanding of the rules that are the foundation of reality: *the laws of nature*.

Everything in life can be explained naturally *because the laws of nature do not allow anything else!* It is the same for the mysteries of life. The explanations can only be natural and simple. Otherwise, they are not true. Faced with these questions, the root of the problem in science has never been the lack of knowledge, but the logical impasses caused by materialistic beliefs—beliefs that are born of erroneous interpretations based on deceptive appearances. It is not necessary to wait for a "great scientific discovery" to solve the enigmas of consciousness and life. Everything is already in our hands. It is only necessary to get out of the dead end of materialism by changing our beliefs, to be able to approach these problems in a new way.

3.3 ARTIFICIAL DIVISION AND ARTIFICIAL CONFUSION

We all tend to cling to our most precious beliefs with an irrational stubbornness, even when we are shown that they are false. This is because our beliefs are, by definition, ideas that we value.

When we adhere to several irreconcilable beliefs that lead us into logical dead ends, but we still want to keep them all because we are attached to them, the solution that will be adopted by our intellect is not to try to reconcile them to avoid disturbing our precious mental comfort with painful questions.

This process is often unconscious, but it is inevitable when we do not want to change our false beliefs. This process has the effect of bringing divisions into our thoughts, which can create false categories in our worldview that exist only in our imagination, false domains that are disconnected from reality. These artificial divisions are intended to allow the intellect to make exceptions to the usual rules when dealing with certain subjects, to protect our most important beliefs.

It is as if, to avoid needing to deal with difficult questions, we decided to believe that in one part of the world "2+2=4," and in another "2+2=5." Of course, this example is a caricature, but even if they are not so obvious, we find everywhere this kind of inconsistency in human thought; it is a fact well known to psychologists, who call this "cognitive dissonance."

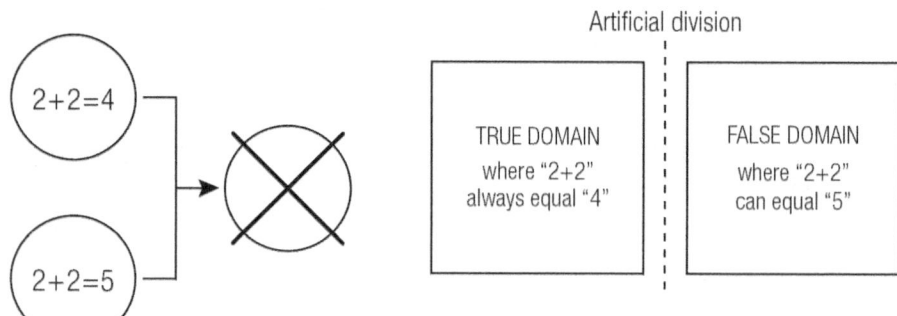

1. Contradictory beliefs lead us into logical impasses.

2. If we want to keep our contradictory beliefs, the intellect installs artificial divisions in our conception of the world. This can create false mental domains, which are disconnected from reality.

When these logical inconsistencies are maintained for a long time, they crystallize, which creates divisions in our minds that may even become dogmas. The most striking example of artificial division is in religions, which support the idea that supernatural phenomena exist, miracles that can escape the laws of nature. Since how these miracles are presented to us is devoid of logical consistency, this type of belief has always attracted skepticism. Confronted with criticism, the explanations of those who believe in miracles generally sound like this: "God is almighty, so he can make all the exceptions he wants." Thus, according to this type of believer, God really can interrupt the normal functioning of the world so that for a moment "2+2=5"! Therefore, in the minds of these people, there is a division between two categories of phenomena: natural phenomena, which obey the laws of nature, and miracles, which are exceptions to these laws. Here is a representation of this way of conceiving the world (next page).

TRUE DOMAIN where all phenomena obey the laws of nature	**FALSE DOMAIN** where certain phenomena escape the laws of nature (miracles)

This belief in the supernatural conflicts with our experience of reality, since we all know that our life is framed from A to Z by the laws of nature, which always manifest themselves with a perfect constancy that is never interrupted. This constancy of the laws is even the basis of science since, without this constancy, science could not establish any rule to explain phenomena and to make predictions. This understanding of the functioning of the laws has led many people to interpret all phenomena considered "miraculous" or "supernatural" as fabrications. However, these people are only jumping to conclusions since what it really means is that there are *necessarily* natural explanations for any phenomenon. In fact, the constancy of the laws of nature means that it is the supernatural interpretation of these phenomena that is false, and not necessarily all the extraordinary events recounted in religious narratives. Some of them may have actually happened, even though they have been misinterpreted later, because the laws of nature offer many possibilities still unknown or misunderstood.

Be that as it may, materialists are not well placed to criticize the fact that religious people believe in miracles since they themselves believe in the existence of phenomena that do not obey natural laws. In other words, they themselves adhere to beliefs that force them to create false divisions in their thinking, and because of that, they need to make strange exceptions when they reason about these subjects.

For example, believing that the brain can create consciousness while admitting that it is of the same nature as the other objects that cannot, requires an artificial division because these two statements are in contradiction. This belief forces the materialists to consider that the brain is in a special category, apart from the other material objects. But, when materialists are asked to explain what is the special attribute of the brain that enables it to produce consciousness while other objects are incapable of it, they begin to stutter. The few materialists who try to answer this question then resort to all kinds of strange explanations, which they do not use in any other case, in the same way that religious people resort to miracles when questions go beyond them.

Similarly, one must install a false division to believe that the first life forms emerged spontaneously from matter, through some mysterious processes relying on chance, at the same time as one believes that it is impossible for all other life forms, which are always the fruit of a reproduction. This belief also obliges the materialists to invent a special category for the phenomena they believe to be at the origin of life, an area in which they use explanations that they themselves would deem absurd in any other context. Indeed, the materialist theories of the origin of life are based on such incredible series of events that adhering to them requires as much faith as believing in a miracle!

We will delve deeper into the inconsistencies of materialist theories in chapters 8 and 9, as well as the solutions offered to us by the universalist approach. For the moment, it is enough to remember that the materialists themselves admit that they are very confused, and that is the reason they consider these questions to be the greatest mysteries of science. Few researchers dare to propose solutions to these problems, and the various theories put forward to date are very far from creating a consensus in the scientific community.

In the course of this book, we will see that the difficulties experienced by the materialists come from the irreconcilable beliefs that they maintain, beliefs that led them into logical impasses, impasses that they call, in a pretentious way, "the great mysteries of science." To keep their cherished beliefs, materialists must reason differently about these so-called mysteries by installing artificial divisions in the same way that religious people use a weird logic when thinking about the so-called miracles. In other words, to answer these questions, the materialists simply replace religious miracles with the "miracles of matter," believing that in very special cases matter can engender consciousness or life, adopting a mysterious behavior completely outside the usual rules.

False divisions are at the root of all problems. That is why we must all try to bring down these walls erected by the intellect. It is this work that we will tackle in this book, regarding the enigmas of consciousness and life, showing that they can find answers in the already known laws of science, without needing to resort to the strange exceptions invented by religions or by materialists.

We are not used to criticizing materialistic thinking in the same way as religious thinking; whereas, in both cases, we find false premises that have the same origin: *artificial divisions invented to protect false beliefs.*

TRUE DOMAIN	FALSE DOMAIN	TRUE DOMAIN	FALSE DOMAIN
where all material objects are unconscious	where certain material objects have the capacity to generate consciousness	where life is always a reproduction	where certain life forms can be generated spontaneously

The worldview of materialists contain artificial divisions that have given birth to false domains where exceptions to the natural laws are permitted. These false domains exist only to protect materialistic beliefs.

All beliefs must be subjected to the same ruthless critical analysis, not giving any preferential treatment to ideas solely because they please us or because they are commonly accepted in our time. There are true notions in all belief systems, but also many falsehoods. Otherwise, there would not be so many contradictions between the different philosophies. We are not here to choose between the various philosophies prevalent in our time, we are only in search of natural and logical answers without worrying about labels.

The main consequence of establishing artificial divisions in our reasoning is the creation of a state of *artificial confusion*. In this state of confusion, many come to believe that the world is incomprehensible; whereas, this sensation of mystery exists only because our intellect is lost in a labyrinth created by our false beliefs. For the "mysteries" to disappear, it is we who must change our beliefs to *adapt* them to reality. There are no mysteries in the functioning of nature since everything obeys simple laws that all can understand.

The root of all problems is that we are more concerned with protecting our false beliefs than with seeking logical solutions. In the same way that a plant that is not adapted to its environment can survive only with an artificial division provided by a greenhouse, our false beliefs also need a closed space to survive because they are not adapted to reality. This closed space is a false category, an artificial mental space in which we follow a logic that is not natural.

These false divisions are harmful because they prevent us from seeing clearly. Therefore, they must all be eliminated. They are not necessary, for it is possible to explain everything without ever making any exceptions to the natural laws.

Our false beliefs are like greenhouse plants. They need to be cut from reality to survive.

4. THE MATERIALIST BELIEVERS

*Scientific thinking stops where
materialistic thinking begins.*

All the problems that come with materialism are because this approach does not bring real explanations, only appearances of explanations. As soon as we choose to adopt this point of view, mountains of problems rise before us, a succession of logical impasses that prevent us from going further in our understanding of nature.

In reality, the so-called mysteries of consciousness and the origin of life are only the great failures of materialism, failures coming from the fact that this philosophy relies on exceptions to the laws of nature. This is the reason the theories of the materialists only work in their *imaginations*. Because, as soon as they try to understand in practice how a material object can generate consciousness, or how life could emerge spontaneously from matter, they encounter a succession of insoluble problems. Unfortunately, rather than wondering if it is their approach that is wrong, they choose to believe that their inability to find solutions is because these are incredibly difficult problems. All this attitude is doing is maintaining false mysteries.

The failure of materialism stems from the fact that this belief system is based on solutions that go against the laws discovered by science, solutions that are not natural. It does not come to the materialist's minds that the many problems that come with their conception of consciousness and the origin of life indicate that these conceptions are false, because the faith they have in their beliefs is too great for them to consider alternatives.

All the confusion that inhabits the materialists comes from the fact that they believe in the appearance that consciousness comes from the brain and in the appearance that life comes from matter. As soon as one ceases to believe these appearances, all the false problems that haunt materialists disappear instantly. There is no reason science should be confused by the question of consciousness and the origin of life since there are many laws discovered by this same science that can resolve these so-called mysteries logically and naturally. To find these solutions, *we must only stop believing that exceptions to the laws of nature are possible!*

There are many similarities that exist between materialistic thought and religious thought. These two belief systems can be very different in certain aspects, but at their hearts, we find the same mental processes. The problems always have the same source: misinterpretations that become false beliefs, beliefs that themselves lead to misinterpretations, forming a vicious circle, a mental prison.

Religions accumulate errors when interpreting ancient texts, while materialists accumulate errors when interpreting scientific data. In both cases, these false interpretations have become beliefs, beliefs that have been organized into systems, that is, into philosophies and ideologies.

In the case of the materialistic philosophy, the main tenets are the beliefs that it is possible for a material object to generate consciousness, and that the first life forms have emerged spontaneously from matter. To adopt such beliefs requires acts of faith because the proofs of the reality of these phenomena are far from being as solid as the materialists claim. These beliefs are essentially based on deceptive appearances.

Materialistic philosophy is no more rational than the different religious philosophies, even if materialistic thinkers make great efforts to pretend otherwise. The purpose of this book is partly to denounce all those evangelists of materialism, who propagate their faith by presenting it under the guise of science. They like to sell us their beliefs by dressing them in a scientific vocabulary, to make them seem more rational than they really are, and all the credibility their ideas have acquired over time depends on this intellectual illusionism. The materialists have played their game so well that it has caused great confusion, to the point that many consider science to be essentially materialistic.

Science is not materialistic; science is *neutral*. Science has no ideology. It is only the set of facts and laws that have been validated by the scientific method. The primary criterion for an explanation to be of scientific value is that it must be consistent with this set of facts and laws. It is to respect this criterion that materialistic explanations must be rejected because they conflict with many laws. While, for its part, the universalist approach agrees in every way with the findings of science, as we shall see.

Scientific thinking and materialistic thinking are two different things, and it is crucial to separate the two if we want to see clearly. Materialistic philosophy is only a belief system, developed mostly in reaction to certain dogmas imposed by religions, dogmas that materialists are often right to reject. But on some issues, they have thrown the baby out with the bathwater, especially by rejecting the possibility that consciousness is independent of the brain and that life can exist in the invisible domains, creating a vision where everything depends on matter. This vision, which gives too much importance to appearances, has led the materialists into logical impasses that they overcome exactly the same way religious people overcome the impasses present in their philosophy: by installing false divisions and by committing acts of faith!

Materialists are far from being as objective as they claim. They are believers, and, like all believers, they deform reality by giving exaggerated importance to the details that seem to confirm their preconceived ideas, and by diminishing the importance of what contradicts them. This mental process is called "confirmation bias," and is a trend that is present in everyone.

People who are too invested in their false beliefs become incapable of objectivity, and if enough of these people are in a position of authority, the situation can become catastrophic because these beliefs will spread under their influence. It is the abundance of false beliefs, and not the lack of knowledge, that has always been the primary obstacle to solving the "mysteries of life." For those who want to progress in their understanding of the world, the most important question to ask is not "What new knowledge should I acquire?" but rather *"What false beliefs should I give up?"*

With time, the beliefs of the materialists have created ever-deeper divisions in their thoughts. As we have seen in Chapter 3, this means that these people reason in a special way when faced with the questions of consciousness and the origin of life, by resorting to solutions that they do not use in any other case. Reasoning using faulty logic is common in religions, particularly regarding miracles, which are by definition events that are considered to be outside the laws of nature. Materialists recognize that this approach is false when it comes to the so-called miracles, without understanding that they use exactly the same approach to bypass the so-called mysteries of consciousness and the origin of life. As soon as they approach these questions, they start reasoning abnormally, permitting all sorts of exceptions only to protect their beliefs. In fact, they only replace faith in religious miracles by faith in the miracles of matter! It is abnormal reasonings of this type that are the source of the artificial confusion that inhabits the materialists, a thick mist they do not know how to eliminate.

Materialistic thought has more in common with religious thought that with scientific thought. This is an affirmation that materialists will fight until their last breath because the belief that their thoughts are more rational than religious thought is one of the fundamental pillars of their philosophy. Everything stems from this pretension, from the idea that they have science on their side—if we topple this pillar, everything collapses.

When one highlights the fact that their philosophy is full of shortcomings, materialists often choose to offer us a counter-argument that can be summarized as follows: "Maybe we do not have answers to these questions, but that does not mean that our philosophy is less useful than another because no one is really able to answer them." In other words, they will bring out this idea so widespread in our time: "We cannot know."

The people who defend this position are what might be called "soft materialists." These people live their everyday lives seeing their consciousness only as a mysterious product of their brain, and considering that life does not exist outside of visible matter. But when you ask these people more in-depth questions about their beliefs, they will admit they are not certain. In other words, they will admit the possibility that consciousness, willpower, and life are maybe something more than the product of reactions between material elements. But for them, it remains only vague possibilities. They prefer the materialistic position because it has always been presented to them as the most rational position, the "default" position of the scientific community.

Soft materialists are not completely closed to other possibilities, but definitely resolving these questions seems to them an impossible task, and that is why these people prefer believing that we cannot know. This idea is, for them, very comfortable. Indeed, why should we make efforts to solve these questions if we can never be sure of the answers anyway? By nourishing this belief, they do not feel guilty about never making the necessary efforts to solve definitively those questions that are, in fact, the most important—those that should be solved *before* anything else.

The behavior of soft materialists resembles, once again, that of religious people. In most cases, religious people follow a religion only because it is the culture in which they grew up, rarely questioning the beliefs and customs inherited from their ancestors. It is the same with most materialists. They have adopted materialistic thinking only because it is widespread in our day, and because that position is supported by people they see as figures of authority, without ever bothering to delve deeper into the absurd implications that come with these beliefs.

We can criticize hard-core materialists, those who believe, for example, that consciousness is useless, that free will is an illusion, and that the existence of life is just an enormous and meaningless accident. But, at least these people have the courage to accept the logical implications that come with their beliefs. This is not the case with soft materialists, who do not want to take any strong position. When one puts these people before the absurd implications of materialism, often they will have doubts about these conclusions which are inevitable as soon as one considers that everything *depends* on matter. Unfortunately, instead of listening to their intuition and reason to permanently reject materialism, most of the time, these people will prefer to go back in the comfortable mist created by the belief that "We cannot know."

It is very pretentious to believe that you have all the answers, but it is just as foolish to voluntarily ignore the solutions we have at hand, only because of the belief that we cannot reach any certainty on the subject of the "mysteries of life." For all these questions, there are definitive answers. But it is not religion that offers these answers, nor materialism, *but the laws of nature*. Everyone can observe these laws to understand how they give us simple and logical answers to the so-called mysteries of life. For that, we only need to follow the threads they offer us, as we will do in the following chapters.

5. THE UNIVERSALITY OF THE LAWS OF NATURE

The universality of the laws of nature is the master key.

In this chapter, and the following ones, we will see the primary tenets in support of universalism. Chapters 5, 6 and 7 will be used to present these premises. Then, in chapters 8, 9 and 10, we will use these elements to build a consistent worldview, in which consciousness and life are no longer mysteries. Without further ado, let's now see the subject of the universality of the laws of nature, the first, and most important, of the tenets supporting the universalist approach.

We need to free science from materialistic beliefs so that it can do its job properly, which is to explain how the world works. The most important when we want to progress in our understanding of reality is not to acquire new knowledge, but rather to be ready to give up our false beliefs since it is these distorted notions that create the so-called mysteries of life. In fact, everyone already has enough knowledge to answer the most profound questions since the solutions are in the laws of nature. And, because we experience the effects of these laws at every moment, we already know them intuitively. So, the only thing that is needed to solve the mysteries of life is to understand the true scope of these laws that we all know.

The laws of nature are the most well-tested facts of science because they come from a very large number of repeated observations, by a very large number of independent researchers. Consequently, those who aspire to base their worldview on science must, first and foremost, be careful to always respect the laws in their reasonings and theories.

Scientists are right to be wary of certain ideas found in religion, pointing out the mistakes those philosophies make when they let beliefs that go directly against the laws take root in them. That being said, we can also conceive that the laws of nature offer many possibilities that science has not yet discovered. Therefore, it is important to maintain a balanced position, and not to reject everything that seems, at first sight, not to fit with science, because it is possible that in deepening the subject we find explanations in accordance with the laws of nature.

It is the same with the idea that consciousness and life exist outside of matter. Materialists often discredit these notions out of hand, by using easy labels to categorize such ideas. For them, those ideas can only exist in the realm of the imaginary, the supernatural or the paranormal, and cannot be part of science or rational thought. This hasty categorization from the materialists allows them to reject ideas that they do not like without much effort. However, contrary to what they claim, prejudices of this kind are only means that they use to protect their beliefs and not positions imposed by science. This is because it is quite possible to explain how consciousness and life can exist outside matter by relying *only* on well-known laws of science, as we will see in this book.

All reasoning about the functioning of the world must be based on the natural laws if it wants to anchor itself in reality. Furthermore, for this foundation to be solid, it is necessary, first of all, to accept the *universality* of the laws of nature.

The principle of the universality of the laws of nature is perfectly simple; it can be defined thus: *The laws of nature act the same way through all of reality.*

The laws are perfectly constant everywhere in the universe; they acted the same in the past and will act the same in the future. In science, this principle is also called the invariance of the laws of physics or the principle of relativity.

The universality of the laws of nature is so important that it can be considered the basis of science, and even the basis of any logical understanding of the world. Indeed, it would be impossible for scientists to develop theories to explain and predict phenomena if the principles underlying these phenomena varied over time or were different from one place to another.

For example, we know that water boils at a temperature of one hundred degrees Celsius, at a pressure of one atmosphere, and crystallizes at a temperature of zero degrees. Likewise, all elements change states at very precise temperatures and pressures—it is a law of nature.

Can we imagine a world where these conditions would continually vary in a random fashion? Something as banal as cooking would become impossible. One could no longer conceive of a recipe since the cooking temperature of the food, and how the ingredients combine, would change continually. It would be impossible to make predictions in such a world; it would be impossible to build something stable, and even life could not develop since everything depends on the fact that the behavior of the elements is always exactly the same when subjected to precise conditions.

As another example, can we imagine what the world would be like if the law of gravity did not manifest itself in a consistent way? One day, we would be light as a feather, and the next, we would be unable to support our weight. The stars would not regularly attract the planets; no stable planetary system could be formed; the universe would be an incomprehensible chaos.

It is the same for all the laws of nature. The heart of the natural laws is their *perfect constancy*, and this regularity is the immutable rock upon which reality is built. If scientists have been able to develop theories and techniques, it is only because the laws that support each of the fields of science never vary, neither in space nor in time.

A law is a rule that an element must necessarily obey in order to belong to a particular category. There are great natural laws, or fundamental laws, that encompass a wide variety of phenomena, and there are smaller laws that apply only to a certain category of phenomena. However, all laws, great and small, are universal in the sense that if a law applies to a certain kind of phenomenon, it must apply to *all* phenomena of the same kind, everywhere in the universe. By definition, a law of nature is a universal principle: if a law applies to an element of type X, it must apply to all of the type X elements of the universe. A law of nature allows no exception in its field of application. Otherwise, it is not a law of nature.

The principle of universality of nature is simple, and everyone already knows it intuitively. What we will see in this book is that this principle is the most powerful of all principles, the one that has the greatest explanatory power. When properly used, it becomes a *master key* that allows us to open all doors and solve all mysteries.

This is because the universality of the laws means that, when one understands the laws governing a particular kind of phenomenon, one can use these laws to explain all phenomena of the same kind—whatever the form in which they occur.

It is this principle that we will use to explain the mysteries of consciousness and the origin of life, showing that these are processes that are similar to others that we already understand very well. The main difference is only that invisible elements are involved, but at the level of the laws, we will see that these processes are nothing special.

However, to understand why the universality of the laws of nature is the key to solving the enigmas of consciousness and life, we must first examine other important concepts. It is essential to progress step by step to avoid tripping. The theory presented in this book is a construction that we will first see piece by piece, before combining these elements to form a vision of reality that is perfectly consistent.

5.1 CONCEPTUAL UNIFICATION

Now, let's see another piece of the solution, which is inseparable from the principle of universality of the laws of nature: conceptual unification.

At first glance, the term "conceptual unification" may sound very abstract, but it is actually a notion that is easily understandable. As the expression says, we make a conceptual unification when we establish links between concepts, that is, when we consider something as part of the same set elements that we considered separately before. It is a mental operation that goes in the opposite direction of the artificial division, which we saw in Chapter 3, and it is through this process that we can remove the false divisions that confuse our thinking.

As an example, imagine that you met someone at an event and that after a long talk with him, this person tells you something that suddenly makes you realize that he is the brother of someone you already know. We all have had similar experiences and have seen what is happening in our minds at the moment when our intellect classifies in a single family those elements that it previously considered separate. What happens then is a conceptual unification.

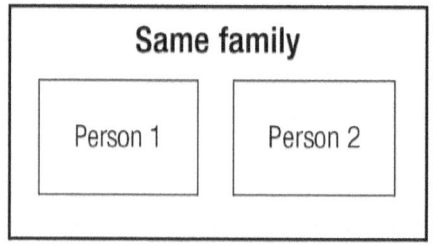

A common example of conceptual unification: understanding that two people belong to the same family.

Our mind is continually manipulating our concepts, our mental representations, using all sorts of criteria to put them into different categories, which can also be called "classes," "groups," "sets," "genres," "types," "domains," or "families." This is the same way we classify objects in boxes or files in folders. When the intellect places concepts in the same domain, this operation is a unification, and if not, it is a division. In both cases, this can be done for good or bad reasons. Therefore, these operations must be carried out carefully, because if we are negligent, our conceptions can easily become disordered, just as is the case in a house if we go too long without doing the housework.

Most of the time, conceptual unification is a mundane operation, but in some particular cases, it is an operation of enormous importance, which enables us to solve issues that were previously considered great mysteries. Throughout the history of science, there are many examples of unification of this kind, which have been events that have changed the way we view the world.

As a first example, there was Isaac Newton in the 17th century, who understood that the force that makes objects fall on Earth is also the one that determines the orbits of the celestial bodies: gravitation. This may seem trivial today, but for the people of the time, this idea that we can use the same mathematical formulas to explain the fall of apples as much as the movement of the moon, was revolutionary. Because, since antiquity, people considered that the laws that reigned on Earth were different from those that reigned in the sky, a false division that lasted for millennia, until Newton elaborated his theory, which unified these two areas.

As a second example, there was James Clerk Maxwell, in the 19th century, who unified electricity and magnetism into a single set of equations, which became the foundation of the theory of electromagnetism. Once again, two phenomena that were thought to have little to do with one another were unified by the same rules. To qualify this unification of revolution is an understatement since countless inventions owe their existence to this understanding of the laws of electromagnetism. By unifying the concepts of electricity and magnetism, Maxwell has opened up a world of possibilities, and he erected one of the pillars of our modern civilization.

Then, there was Albert Einstein, at the beginning of the 20th century, who, with the notion of space-time, unified the concepts of space and time in his theories. This genius also gave us the most famous example of unification, with the formula $E=mc^2$, which establishes the link between mass and energy.

In fact, as an example of conceptual unification, one could include all the equations of physics. These equations are forms of unification since they tell us that the terms located on each side of the equal sign (=) are equivalent concepts, that they are two different ways of speaking of the same thing.

One might think that advances in physics are being made by discovering more and more laws, but these examples show us that, instead, it is the opposite: real progress is made when we can explain more and more phenomena with *fewer and fewer* laws. Whenever we succeed in explaining with the same laws phenomena that we explained with different laws in the past, it is an important conceptual unification and another application of the principle of the universality of the laws. Since this approach has been so successful in the past, physicists continue in this direction even to this day, attempting to unify all fundamental concepts of physics into a single set of rules, sometimes called the "theory of everything."

This way of progressing in our understanding of the world, by unifying concepts, applies not only to physics but to all domains. For example, in chemistry, the most important advance occurred when researchers realized that the different elements of nature were all

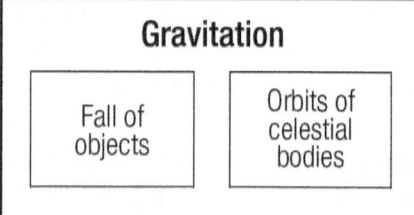

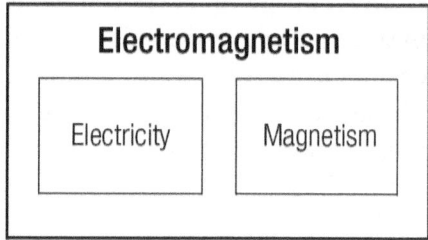

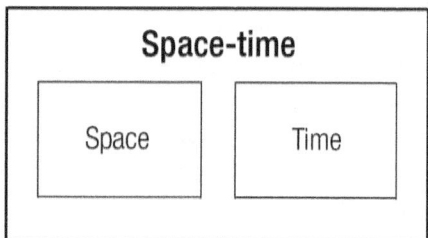

Examples of conceptual unifications, which are among the most important steps in the history of science.

composed of the same sub-elements: protons, neutrons and electrons. Whereas, previously, the elements were classified in more or less arbitrary categories, this conceptual unification allowed chemists to classify the elements in a much more rigorous way, by the number of protons their nuclei contain, a classification that became the periodic table of the elements.

In the field of biology, another famous example of unification is the development of Charles Darwin's theory of evolution by natural selection, which originated when he understood that the principle of transformation by selection, by which a breeder can modify a species by selecting the breeding animals, must also apply to animals in the wild.

Also, in biology, it was long believed that there were two types of life: life that comes from a reproduction process and life that emerges by spontaneous generation. Indeed, it was once thought that certain life forms could appear spontaneously when particular conditions were met, be it microbes in a liquid, maggots in rotten meat, mice in a pile of straw, and so on. These superstitions disappeared with the development of science because, throughout their research, scientists have gradually discovered that in all cases where people believed in spontaneous generation, life had indeed been transmitted from elsewhere, only in a way that was misunderstood before. Finally, scientists had to face the fact that spontaneous generation is only an imaginary phenomenon, and that all life forms belong to the same category: life that comes from a reproduction process. Moreover, this extremely important conceptual unification is not quite complete today, since the materialists persist in believing that some form of spontaneous generation was possible for the first life forms that appeared on Earth, a subject that we will discuss in detail in Chapter 9.

There are many other examples of unifications throughout the history of science, one could write a book on this subject alone. What must be remembered is that these conceptual unifications were possible because those who made them *went beyond the divisions maintained by their contemporaries*. These researchers made important steps in the progression of knowledge, understanding that certain phenomena that seemed to be separated from a certain point of view are actually fundamentally linked. For that, they had to be ready to cross certain limits, to go beyond the artificial divisions that their contemporaries maintained in their conceptions of the world.

Nothing is more important for the one who wants to understand the world than to seek those unifications, because it is only through this process that we can build an ever more coherent vision of the world. The junctions that we accomplish this way enable us to reach a higher level of understanding, where we realize that what previously seemed to be separate elements are actually different manifestations of more fundamental principles, like different branches that come from the same trunk. Rising to ever-higher concepts, we arrive at the main trunk of reality, to the concepts that encompass all other concepts: the laws of nature. These laws are the highest concepts, since all the other concepts, all the elements, all the phenomena, are only *fruits* of the laws of nature, consequences of their existence. The laws of nature are the concepts on which all other concepts depend, and therefore, those that possess the greatest explanatory power. When they are well understood, these laws can explain everything.

If we want to understand reality at its core, if we want to solve the biggest questions, we must always think using the highest concepts, which are the natural laws. A task that may seem daunting, but that is actually easy since the most important laws are also the simplest!

It is like rising from the surface of the Earth to go into space. As we ascend, the great diversity of objects that exist at the ground level merge gradually into an increasingly uniform landscape, which then transforms into a sphere, which itself becomes a luminous point, the simplest form there is. It is the same in the domain of thoughts: *The higher our understanding is, the more our vision of the world becomes simple.*

The ideal of science is to explain all phenomena with the same laws. The conceptual unifications that have taken place throughout the history of science have accomplished some of this task, but to continue further, it is necessary to abandon the artificial divisions created by materialism.

What we need to look for is an increasingly unified vision of nature, a vision in which the same laws apply everywhere, including in the realms of consciousness and life. *To obtain this unified vision of reality is the goal of universalism. Nothing else interests us.*

This goal may seem very pretentious, but we will see that it is not nearly as difficult as one might think. In fact, much of the difficulty arises only because we are used to believing that the enigmas of consciousness and the origin of life are extremely complicated problems; when, in reality, they are easy to solve when they are approached in the right way, that is, when we use an approach based on natural laws and not superficial appearances. This belief that these are difficult problems confuses us because if we approach a question with the prejudice that the answer must be complicated, and then we are presented with a simple solution, we risk rejecting it, believing that it is not what we are looking for! Only to get lost after in a labyrinth of useless complications that we deem more worthy of our intelligence... Again, we must not underestimate how our prejudices and our beliefs can confuse our reasonings, and, among all mental parasites, the belief that the answers to the mysteries of life must be difficult is undoubtedly one of the most common and most harmful!

All the solutions that are used by universalism are very simple. Not only are they simple, but they are based on laws that are already well known to science. There is nothing new in these concepts, only the way they are used is different. In other words, it is about taking already known elements and reorganizing them, reinterpreting them, classifying them differently, by establishing logical links that are different from what we usually see.

The heart of the problem concerning the great questions of existence is not that we lack the knowledge needed to answer it. We already have all the necessary knowledge; the problem is only that these elements are poorly organized in our heads because of our false beliefs! Many logical relationships are in the wrong place, many concepts are classified in the wrong categories, and even many of these categories exist only in our imagination... In short, it is a mess! It is a disorder that is mainly caused by materialistic beliefs, which are wreaking havoc in our day and age.

The theory presented in this book is essentially about reorganizing the concepts that are so badly placed in materialist theories. For example, we saw in Chapter 3 that materialists place the brain in a category apart from other material objects, considering that it possesses the extraordinary power of generating consciousness. This false categorization is the source of great confusion for materialists because when they try to understand where this special power comes from, they cannot. It is a failure of their theory that they present to us as a "great mystery."

Within the universalist approach, this so-called mystery does not exist since the brain is considered like all other material objects, namely, as an *unconscious object*. For universalism, all that is material must be considered in the same way: these are non-conscious things, non-sentient, not endowed with sensation, things that do not have the capacity to perceive or feel... All that is material is unconscious; it is a law of nature.

As a material thing, the brain has no particular power. It is an object as unconscious as the others; an object that, like all the others, can only act as the *intermediary* of a conscious activity. Therefore, within the universalist theory, this organ is classified in the same category as the other material things, namely the category of unconscious objects, and this conceptual unification eliminates the "great enigma of the brain," which exists only in the imagination of materialists.

Similarly, we have seen that materialists place the first life forms in a particular category, believing that they have appeared spontaneously through mysterious processes relying on chance, a false category that also creates all sorts of problems that mystify materialists when they try to understand how such a miracle could have occurred.

Once again, within the universalist approach, this mystery is nonexistent, since the first life forms are considered in the same way as all other life forms, that is, as *reproductions*, a conceptual unification similar to the one that led to the abandonment of the belief in spontaneous generation which we saw earlier.

To remove these mysteries, we must stop making strange exceptions that create false divisions in our thoughts. Some key concepts need to be reorganized, considering all objects as unconscious, even the brain, and all life forms as reproductions, even the first life forms:

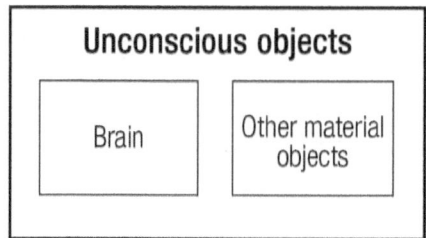
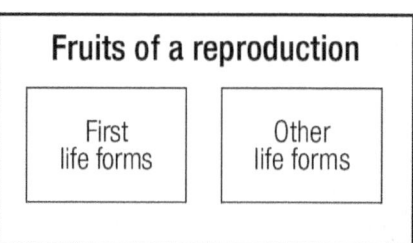

By including the brain in the "unconscious objects" category, it is united to the other material objects. Likewise, when the first life forms are included in the category "reproductions," they are united with other life forms. The brain becomes only a particular case of unconscious object, and the first life forms a particular case of reproduction, phenomena that can then be explained with the same laws as the other phenomena of their respective genre, laws that we already know very well.

In this book, we will see why mental operations of this type, which are the simplest, are also the most *powerful*. Used in the right way, they can solve the greatest mysteries of existence. Conceptual unifications are at the heart of the greatest theories of science, and they are the key to solving the mysteries of consciousness and life.

As we saw in Chapter 2, understanding the world is *only* organizing the concepts correctly! To see things clearly, that is what we need to focus on—the rest is just details.

In chapters 8 and 9, we will see in detail why the brain is necessarily an unconscious object, and the first life forms necessarily reproductions, seeing as this is the only explanation that fits with the laws of nature. For now, it is normal for these statements to give rise to several questions because they are only pieces of the solution. We must see other notions and other conceptual unifications before our portrait of these phenomena is complete. In particular, we must first understand the crucial role of the invisible, a subject that we will see in the next chapter. Another piece of the solution is the understanding that consciousness and life have their sources on the invisible side of reality. We will see that, contrary to widespread beliefs, this invisible domain is not esoteric since the discoveries of science, as well as the principle of universality of the laws, allow us to have a good understanding of it.

Consciousness and the origin of life are phenomena that can be easily understood, even if they imply the existence of invisible realities, since they follow the same laws as phenomena that are well known. It cannot be otherwise, because since these laws are universal, they encompass the visible domains as much as the invisible domains.

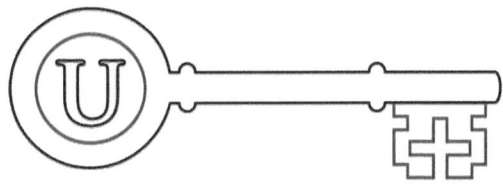

The universality of the laws of nature is the master key, the one that opens all doors. It allows us to understand the infinitely small as much as the infinitely large, the past as much as the future, the visible as much as the invisible.

6. THE IMPORTANCE OF THE INVISIBLE

Humanity only knows the shadow of reality, the essential is invisible.

The invisible part of nature is much more important than its visible part. This is one of the primary lessons of science.

Whenever scientists have access to new tools that allow them to see realities that were invisible to us in the past, what they discover often exceeds all their expectations. For example, the invention of the microscope allowed us to discover the world of microorganisms, millions of amazing species, an extremely rich previously unknown life. This life, invisible to the naked eye, was also the key to answer many questions. In particular, it explained several diseases and allowed medicine to take a big step forward.

In another area, the invention of the telescope also allowed us to discover worlds of infinite diversity, enabling us to see planets, stars and galaxies that were invisible to us before. As soon as scientists were able to look at these unexplored areas, what they found exceeded their previous conceptions. Even the best theories do not allow us to conceive the true richness of nature, which does not know any of the limits of our imagination.

Significant scientific progress is made every time we can enter a field that was previously invisible to us, which is why scientists are continually working on designing new instruments that can probe the invisible. It is still the case today with particle accelerators and the other advanced detectors used by physicists, which have allowed us to confirm the existence of many particles previously invisible, without which it is impossible to explain how the world works.

The history of science can be seen as a great adventure, and many of its pioneers have been explorers of the invisible, who have realized that the key to certain puzzles must be on the hidden side of reality, beyond the limits of the instruments of their time. It is this path that we will also follow, turning toward the invisible to explain consciousness and the origin of life. Indeed, since similar approaches have been so successful in the past, why not continue in the same direction?

Often, in the past, when scientists were unable to find satisfactory explanations in visible domains, they turned to the invisible to find answers. The best examples of this are in physics, where many particles were first conceived purely theoretically, as invisible solutions to certain problems, only to be directly observed decades later.

It is the same thing for the enigmas of consciousness and life, *they indicate to us that we must again turn to the invisible,* a solution that is neglected by most current theorists, even if the history of science continually reminds us that neglecting the importance of the invisible is a big mistake!

This disregard is due to the influence of the materialists, who maintain that the visible is sufficient to explain consciousness and the origin of life. Materialist theorists put the visible in the center of their theories, but it is only a preference, a bias that is not imposed by science, but rather by their beliefs. This because the "visible" is only what is possible to measure with the instruments of a given time. It is a category created by the limits of our senses and our instruments, which only have access to a tiny part of reality, and giving too much importance to the phenomena that are part of this

Examples of substances that were initially invisible solutions
THE NEUTRON: Predicted by Ernest Rutherford, to explain the difference between the atomic number of an element and its atomic mass.
THE POSITRON: Predicted by Paul Dirac, as a logical consequence of his equation describing the electron.
THE NEUTRINO: Predicted by Wolfgang Pauli, to explain where the missing energy goes in the beta decay process.
THE GLUON, THE W$^+$, W$^-$ AND Z BOSONS: These particles, proposed as vectors of nuclear interactions, were first conceived theoretically before they were observed in particle colliders.
THE HIGGS BOSON: This particle was first predicted by Peter Higgs, François Englert and Robert Brout. Its existence then became essential to explain how particles acquire their mass.

small category, such as the materialists do, can only lead us into dead ends. The visible is a category the limits of which are always changing. Because of this, theories that rely exclusively on the visible are like houses built on quicksand: they risk collapsing each time a discovery pushes back the limits of our knowledge.

Using invisible solutions is quite permissible by science because all that science requires is that we must have *good reasons* to do so! And, as we will see, there are excellent reasons to turn to the invisible side of nature to explain the mysteries of consciousness and life.

6.1 THE INVISIBLE SIDE OF NATURE

The discoveries of science have long since established that our senses perceive only a small slice of reality: Our eyes perceive only a tiny part of the light spectrum, our ears hear only a tiny fraction of the range of sounds, our sense of smell perceives very little of the odors that exist... This is known to most people. We all recognize that, for example, some animals can perceive colors, sounds, and smells to which we have no access. This limitation of our senses is well known, but the discoveries of science allow us to go much further, to show that what we perceive is only a small part of the phenomena that actually occur around us, and that reality is essentially invisible.

Let us take as an example subjects that are becoming increasingly important in astronomy and physics: dark matter and dark energy. Those who see these terms for the first time may think they come from the world of science fiction, but this is not the case because they have been coined by physicists. The terms "dark matter" and "dark energy" may seem strange, but they are only used by convention and could have been different. In the case of dark matter, we could also use the term "invisible matter" or "matter of an unknown kind," and it is the same for dark energy. It is simply a kind of energy that is not well understood.

The knowledge of these invisible domains has developed gradually over the last few decades, thanks to ever more accurate observations of the universe by astronomers. The discovery of dark matter comes, among other things, from the observation of the rotational speed of stars around the center of galaxies, which cannot be explained without the presence of invisible matter in proportions much more abundant than known matter.

While in the case of dark energy, it has been theorized to explain the acceleration of the expansion of the universe, which requires the presence of a phenomenal amount of energy of an unknown kind.

We will not go into the details of the history of these discoveries here. What is most important to remember is this conclusion: According to astronomers' current approximations, ordinary matter and energy form only about 5% of the mass of the universe, the rest—95% of the universe!—consists of unknown realities, about 27% of dark matter and 68% of dark energy. In addition to this finding, we can reasonably assume that these enormous domains of matter and energy are not uniform, but consist of several different kinds of substances, substances that can be inhabited by as great a variety of phenomena as what we observe in the visible universe. In short, we know almost nothing about the universe...

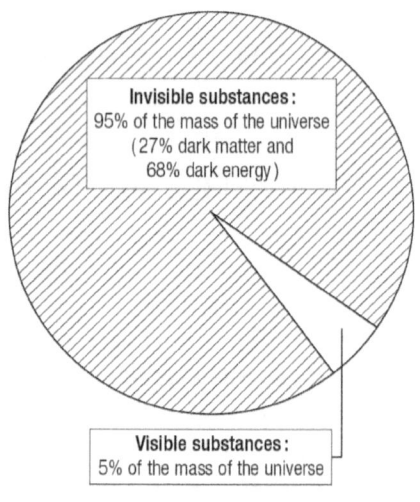

Dark matter and dark energy are just examples of invisible substances that are part of the landscape of contemporary physics. In fact, *there is no known limit to the number of different particles that can exist.* This is why theorists are always ready to conceive of new types of particles when they have good reasons to do so. Another example of this is that of string theory, which states that for each known particle type, there would be another particle, which is its "superpartner," a kind of complementary particle, which has never been observed.

These unknown forms of matter and energy are good examples of the importance of the invisible in scientific theories, but that's not all because we must also include in the hidden side of nature the fact that even *known* particles can combine their activity to form phenomena that escape our senses. Phenomena that the instruments of modern science have not yet allowed us to discover because each type of detector is specialized and limited in various ways, so they cannot see everything.

Scientists will *always* need to include invisible elements in their theories because the instruments of science will *always* be limited. We must accept this fact because disregarding invisible solutions can only produce a worldview that is disconnected from reality. To claim to know nature by considering only what is visible is like claiming to know the ocean by considering only its surface!

Many researchers doubt the existence of dark matter, as well as the many other hypothetical particles proposed by some physicists. But what is essential here is not knowing exactly which of these particles actually exist. What is important is to understand *it is allowed to use invisible solutions to answer scientific questions.*

Concerning the enigmas of consciousness and life, materialists have a preference for solutions that are in the visible realm, while the universalist approach considers that the solutions are in the invisible realm. Invisible solutions have a bad reputation in scientific circles because they are often linked to religion, the paranormal, and the supernatural. It is true that humanity has always had a tendency to fill the invisible domains with inventions of all kinds. But to consider all invisible solutions this way is false. It is, in fact, a very harmful false unification.

In the preceding chapters, we have spoken of false divisions and the confusion they may cause in our thoughts. It is the same for the false unifications that occurs when one classifies in the same category elements that do not go together. This is what materialists do when they consider invisible solutions only as magical thinking.

From a scientific point of view, for the use of invisible solutions to be acceptable, it is sufficient that the existence of invisible realities is a *possibility*. As we have seen in Chapter 2, good use of the scientific method *requires* that we consider all possibilities, even those we believe unlikely. This means that considering invisible solutions to the enigmas of consciousness and life is an exercise in scientific rigor since it allows us to explore all possibilities, even those that are generally neglected. The history of science has repeatedly shown that solutions are sometimes invisible, and it is quite possible that it is also the case here. In fact, the only good reason not to explore invisible solutions would be if the visible solutions had proved to be 100% satisfactory, which is obviously not the case. Otherwise, we would not consider these subjects the greatest mysteries of science!

No matter how one analyzes the question, the conclusion is always the same: *It is allowed to use invisible solutions to solve the mysteries of consciousness and the origin of life.* If it is rational to use the invisible to solve certain mysteries of the universe, it is equally rational to use the invisible to answer these questions. This does not go against science, only against certain beliefs.

6.2 THE LAW OF SELECTION

The more science progresses, the more it realizes that the invisible part of nature is much more important than previously thought. This fact alone should cast doubts in the minds of those who try to explain consciousness and the origin of life relying only on the part of reality that is visible to us.

It is clear that the existence of invisible worlds does not go against the discoveries of science. On the contrary, many indirect observations tell us that most of nature is invisible. Perhaps many still see the existence of invisible substances as a concept that is strange and hard to understand. Let them be reassured because the principle that makes their existence possible is, in reality, very simple, so simple that it can be summarized in one sentence: *Every interaction is selective.* This principle is an important law of nature, which we will call the *law of selection.*

Each interaction is possible only when precise conditions are met, and if we do not perceive the invisible parts of nature, it is simply because we do not fulfill the necessary conditions to interact with them. The central concept that explains the existence of invisible worlds is that of *interaction*, and understanding how interactions govern nature is an essential key to understanding the structure and functioning of the universe.

Our experience of reality comes solely from the interactions we have with it, and this has implications that go far beyond what is generally believed. For example, we tend to think that what we can touch with our hands has more consistency than the images we perceive with our eyes, that it is somehow more "real." However, is that the case? The answer is no because, according to what physics tells us, the way our hands perceive objects is similar to the way our eyes perceive images.

This is because the electrons on the surface of our body, and those on the surface of the objects we are in contact with, repel with such force that they always keep a distance. The result is that we cannot really touch anything because the matter of our hand never really reaches the matter of the object. There is always a space between them—a space that, relative to the dimensions of the particles, is enormous.

Alternatively, even if we are not really able to touch the object that is in our hands, between us, there is an exchange of impulses by means of electromagnetic waves, an exchange of particles of light, and it is this transmission of energy that is perceived by our nerves. So, our hands only capture light-transmitted energy, like our eyes, and similarly for all the senses. It is these different streams of information that our senses have captured that are then united to form an *image* of reality.

What we perceive from the world through our senses is *never* matter itself, it is always only this incredible diversity of waves that combine to generate images. These images have more dimensions than those of photography or cinema, but they are still images since they are formed according to the same principle. What must be kept in mind is that this image of the reality we perceive *contains only the information that we have managed to capture,* and this information is only a tiny fraction of the information that is really present around us. This image is a thin slice of reality.

This phenomenon is comparable to the functioning of the television, the radio, or the Internet; in these cases, we also perceive at every moment only a very small part of the information that is offered to us. For example, if we choose to watch a particular television show, that does not mean that the others suddenly cease to exist, the only difference is that we do not *interact* with these other programs, and from our viewpoint, the result is the same as if they did not exist, but, of course, it is only an illusion. So, we can compare our senses to antennas, relay centers that operate a strict selection, as if we were forced to watch only one TV channel, not having access to the hundreds of others.

What physics tells us is that we can also apply this same principle to the sense of touch, which is limited like the other senses, with access to only a small part of reality. There may be forms around us that are as concrete as what we can touch with our hands, but that the matter of our body cannot touch only because it does not have the ability to interact with them. Palpability is something *relative,* what is palpable at one level, is impalpable at another level, and vice versa.

This affirmation will surprise many, but the principle that makes this possible lies at the heart of nature. It is the law of selection: Every interaction is selective.

An image that illustrates how this law works is that of a key and lock system. For an interaction to occur, precise conditions must always be fulfilled. In other words, the right "key" must meet the right "lock." It is only this that determines whether the connections are possible, and in nature, these complementary elements—these keys and these locks—can take an infinite variety of forms.

For example, the key can be a grain of pollen, and the lock a flower of the right species ready to be fertilized; the key can be a seed, and the lock a soil adapted to this type of seed; the key can be a protein, and the lock the right sensor on the surface of a cell; the key may be an atom that possesses too many electrons, and the lock an atom lacking electrons; and so on...

Everyone can observe this principle of selective interaction at work everywhere, if only in an example as banal as the fact that left hands only go in left gloves! Each can complete with their own observations, to understand by themself the importance of this principle. In this book, it is especially in the field of physics that we will deepen this principle, to explain how it allows the existence of invisible worlds.

In physics, we can consider that the keys are the *force fields* while the locks are the elements *sensitive* to these forces. We can illustrate this with the help of a well-known example of a force field: a magnetic field. You can be in the presence of a magnet powerful enough to lift a car but still pass near it without being lifted yourself. This is because the magnetic interaction is selective: If you have nothing on you that is sensitive to magnetism, nothing happens. In this case, at a certain level, we can say that for this magnet you are invisible.

Another more spectacular example of this principle is the neutrino, a particle present in abundance in the universe, but that interacts very little with the other forms of matter, which means that a neutrino can traverse through something as huge as *the entire Earth* without being deviated from its trajectory. To use an image, it means that if we could travel in space aboard a ship made of neutrinos, we could pass through a planet without even noticing it. For us, it would be only *empty space*. This, only because the material of which our ship is made does not have the capacity to interact with the matter of the planet we encounter on our way. In this case, we can say that for neutrinos, the Earth is invisible.

Nothing is visible or invisible in itself. Nothing is palpable or impalpable in itself. Nothing is penetrable or impenetrable in itself. *It is only a question of interaction.* In other words, those are only *relative* notions. If the interaction is possible with us, the element is visible for us. Otherwise, it is invisible, and that's it! When we speak of the invisible part of nature, or of hidden worlds, it is only in relation to our viewpoint, our senses, and our instruments. As soon as we have the means to perceive them, these invisible levels of reality appear to us filled with visible forms, just as tangible as those we are used to perceiving. In the same way, what is usually visible to us can become completely invisible if we change our viewpoint. This shows us how our experience of reality is built exclusively from our interactions with the world around us, hence, the importance of understanding how interactions work in nature.

Earth from our point of view

The whole universe is structured by interactions because each force of nature, also called *fundamental interaction,* has its own field of action and acts only within it. Physics is centered on the study of these fundamental interactions, and scientists have discovered four of these forces so far: gravitational, electromagnetic, weak nuclear, and strong nuclear.

Earth from the point of view of neutrinos

Each of these fundamental interactions has unique features, but, for simplicity, we can conceive that the action of all these forces is by means of fields comparable to a magnetic field. A force field can be visualized as a network of lines of force. These force fields are everywhere; they permeate every part of the universe, directing all phenomena.

Within a force field, a particle can react in three ways: Either it is attracted by it, it is repulsed by it, or it is indifferent. The third option, indifference, may seem at first sight without much interest, and may even leave us indifferent, but it would be a mistake not to reflect on this option because it is this one that allows physicists to explain the existence of invisible worlds! Inside a force field, a particle indifferent to this kind of force will continue its path without deviating from its trajectory *as if this force field was nonexistent.* In other words, elements that are indifferent to each other are *invisible* to each other.

This simple fact, apparently insignificant, opens worlds of possibilities. It is the possibility of being indifferent to each other, that is, of not being sensitive to the same forces, that allows particles to exist side by side without interacting; and this possibility allows us to understand how an extensive variety of substances and activities can exist in the same environment without interfering with one another.

Let us go back to the theory of dark matter, which tells us that all the celestial bodies, even the Earth, have an environment of unknown particles in a much greater proportion than those that are known. At first glance, this may seem like an extravagant and hard to believe concept. Yet, the principle that makes this possible is of extreme simplicity: To make this possible, it suffices that these substances are *indifferent* to the forces that organize known matter! This observation also implies that these unknown particles must be organized by forces that are also of unknown kinds, forces to which the visible particles are indifferent.

For the theories of physics, a force field is a network in which circulates a particular type of particle, nicely called "mediator bosons" or "gauge bosons." The particles of light, the photons, are bosons of this type, responsible for the electromagnetic interaction, a force that links the electrons to the nuclei of atoms as well as the atoms between them. Likewise, each force has its own mediating particles that act similarly to the photon. The fact that the action of a force implies an exchange of particles is

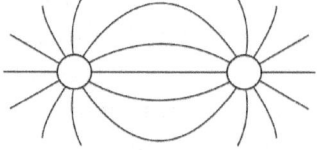

Attraction

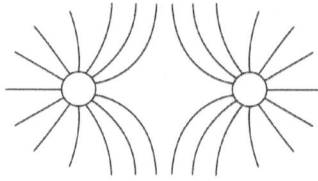

Repulsion

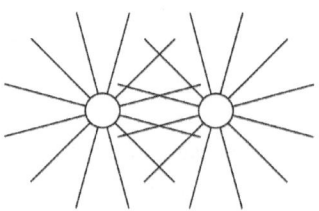

Indifference (invisibility)

Interactions between particles happen through force fields, of which there are several different kinds. Particles that are not sensitive to the same forces are invisible to each other.

the reason why, in physics, the word "interaction" is often used instead of the word "force."

Since particle exchanges are at the heart of the action of a force, it means that when we discover new particles, we also have the opportunity to discover new types of forces or interactions. What this means, above all, is that we have not yet found a limit to the number of different forces that can exist in nature, as we have not yet found a limit to the number of different particles that can exist. The theories of physics offer us no reason to believe that the particles and forces discovered so far constitute everything that exists. On the contrary, everything indicates that a lot more is yet to be discovered!

The existence of other particles and other forces can be easily conceived, the problems that limit our knowledge of them are mainly of a technical nature. Indeed, since we use instruments made of materials obeying certain forces, *we cannot use these instruments to detect substances indifferent to these forces.* In other words, we can theoretically conceive the existence of numerous new particles obeying new forces, but detecting them directly is impossible with the help of instruments that do not obey these forces.

This observation might lead us to believe that the existence of these invisible domains is doomed to remain speculative. However, this is not the case since it remains possible to prove their existence indirectly, thanks to a good understanding of the laws of nature. For example, this difficulty has not prevented astronomers from making many observations that point to the existence of a large quantity of unknown substances, thanks to gravitation. Indeed, among all the forces, gravitation is particular since it acts on all the substances, but in an extremely weak way. Gravity cannot be used to detect particles individually, but we can use it on a galactic scale to detect the gravitational effects of all types of particles, regardless of whether they are visible or invisible substances. This is how the researchers came to the conclusion that the universe is primarily made of unknown realities, dark matter and dark, energy since many gravitational observations cannot be explained without the presence of these elements. Universalism also considers the existence of consciousness and of life as indirect proof that there are important invisible realities, a conclusion based on the principle of universality of the laws of nature, as we will see later.

The law of selection explains why physicists have no difficulty conceiving the existence of invisible worlds, inhabited by structures as concrete as those around us, but which we cannot perceive with our senses or our instruments because the substances they are made of do not have the ability to interact directly with these substances of different kinds. This is natural when one understands how the universe is organized solely according to interactions that each has strict boundaries.

Instead of seeing the invisible worlds as something strange, supernatural, esoteric, or paranormal, one must learn to recognize their existence as the most *natural* thing. The richness of nature is absolutely inconceivable, and this diversity is the reason we are only allowed to interact with a tiny fraction of the phenomena that surround us. We would not be able to function if we were not limited this way because the abundance of information we would receive would make us completely crazy. It would be like watching every TV channel at the same time! It is not surprising, then, that nature has limited our perceptions so much; it is, in fact, a kind of protection.

The universe is structured by a vast network in which waves circulate, waves that exist in a great variety, and each element is sensitive only to part of those waves. This scale of interactions is analogous to a range of frequencies since it allows activities, phenomena, structures, forms, realities, to exist in the same space without interfering, such as radio broadcasts can exist in the same space without interfering. In short, the law of selection leads to the creation of *levels* in nature. These levels are worlds just as rich and diverse as the one that is visible to us, but about which current science ignores almost everything.

So, what is so weird about considering the existence of invisible substances that can serve as a support for consciousness and life, if it allows us to explain observations that cannot be explained otherwise? The laws of physics explain simply how invisible realities can exist, and the progress of science has always confirmed the importance of the invisible. So, if we listen to what the discoveries of science tell us, everything indicates that this is the right direction to follow!

These reasons are already sufficient to give the invisible a lot more significance in our worldview, but we will see another one. Again, this reason is the most important of all principles: the universality of the laws of nature.

6.3 THE INVISIBLE AND THE UNIVERSALITY OF THE LAWS OF NATURE

Even the most skeptical, if they study science seriously, must admit that the existence of invisible realities is entirely in accordance with the laws of physics. But these people will say that, even if these unseen worlds exist, we cannot turn to them to find solutions since we do not know what they contain.

This attitude is false because even if we cannot know in detail what exists in the invisible domains, we can still understand some essential aspects about them using the universality of the laws of nature. Indeed, even if we do not know much about the invisible worlds, we know at least one thing with certainty: *The laws of nature must act in the invisible domains the same way they do in the visible domains.*

The universality of the laws must apply in all cases, even in the invisible. We must constantly keep in mind that this boundary between the visible and the invisible, which seems so important to us, is created only by the limitations of our senses and instruments. From the viewpoint of the laws of nature, this limit does not exist, since the laws treat all things in precisely the same way, both what is within this limit and what is outside.

It is a big mistake to believe that the natural laws, which we see everywhere in action, suddenly stop at the limit of our senses and instruments and that beyond this horizon, everything becomes strange and incomprehensible. One just needs to think about it for a few minutes to understand how absurd this position is—and yet, it is a belief that is maintained by most of humanity! This, as much by religious people, who believe that the invisible is a domain outside natural laws, as by materialists, who consider that we cannot know anything about the invisible, and therefore, that it is a domain that we must not care about.

Those attitudes are false since it is a certainty that the universality of the laws of nature must be applied in both the visible and the invisible, and this principle allows us to answer many questions that one can ask about the invisible worlds.

What this principle tells us is that the invisible is not a rupture of the natural order, but a part of the natural order. This means that nature, as we see it around us, does not become entirely different when we enter the invisible realms because the laws always act in the same way.

Despite the incredible diversity of phenomena found in the universe, the basic structure of nature remains the same everywhere. This basic structure is so simple that it can be summed up in one word: *levels*. Nature is made of levels, layers, degrees, zones...

This natural order, everyone knows it very well, since there is nothing else. Wherever you look, you can only find levels! The ground on which one walks, the atmosphere one breathes, the oceans and their depths, or the biosphere made up of countless creatures, all these are just different levels of nature. Starting from the level of particles, then atoms and molecules, which combine to form various elements, which themselves combine to form planets and stars, which then form stellar systems and galaxies... Nature is a succession of degrees, which themselves are composed of very numerous sub-degrees, up to their finest details.

Our body is also composed in this way. Starting from the skeleton, on which we add layers of muscles, nerves, and blood vessels, on which we add the last layer, which is the skin. And of course, each of these layers is composed of sublayers, as is the case for the skin, which is composed of the epidermis, the dermis, and the hypodermis.

Even artificial structures, formed by humans, are composed of a succession of levels since the same laws apply. For example, a house is comprised of various layers, which are floors, walls, ceilings, and roofs. The clothes we wear are made of layers of fabrics, the sandwiches we eat are made of layers of food, and the cars we drive are made of layers of metals and other materials. Even the book you are reading is made of successive layers, which are the pages of the book, as well as its cover, and, if you read it on a screen, it is also a layer composed of pixels...

We could add examples indefinitely since *everything* that exists is composed of levels! This natural order is governed by the law of selection, which tells us that all interaction is selective. This great law pushes the compatible elements to combine, which separates the elements that go well together from the other elements with which they are less compatible, that is to say, those with which the interaction is weaker, or with which there is repulsion or indifference. This phenomenon leads to the creation of zones and levels, within which only compatible elements coexist; it is a simple process that everyone can see at work everywhere. From the extremely small to the extremely large, nature is an immense structure made of successive levels. Within each of these degrees, a great variety of compatible phenomena coexist, having between them exchanges that are necessary to maintain a particular state of equilibrium.

Nature is made of levels. It is an observation that seems insignificant, and yet, this ordinary idea is sufficient to understand some crucial aspects of the invisible domains. Because the law of selection allows us to answer, in broad strokes, the question, "What does the invisible contain?" The answer is obvious: *levels!* The invisible is composed of levels, each filled with a great variety of phenomena, just as the visible is composed of levels, each filled with a great variety of phenomena. The answer cannot be anything else because, otherwise, it means that the laws of nature are not universal, which is absolutely excluded!

The invisible is composed of levels, just as the visible is composed of levels. This is a fundamental conceptual unification, which everyone can use to understand the world.

This unification allows us to have an *overview* of what the invisible contains, but obviously, it does not allow us to know in detail what the various invisible degrees of nature contain. But these details are secondary, since an overview is sufficient to answer the big questions, as we will see. We must avoid getting lost in the details because they are useless for the questions that occupy us, and because the details of nature are inexhaustible.

For example, it is possible to spend a lifetime studying a *single* living cell, and always discover new phenomena! The smallest blade of grass contains a variety of elements beyond our understanding; each level of nature is of an inconceivable richness. All the details of nature are fascinating, but they can also be a trap because they can make us lose sight of the big picture, lose sight of the overview, which alone can give us the answers we seek concerning consciousness and life.

Although we will not be concerned about the details in this book, it is still essential to understand that the invisible levels of nature contain just as much diversity as the visible levels because they are formed by the same laws that give birth, everywhere in nature, to an infinite variety of phenomena. So, here is another conceptual unification, essential for building a coherent vision of reality: *Invisible levels are very diversified domains, just as visible levels are very diversified domains.*

Just like the visible, the invisible is made up of levels, each filled with a great variety of phenomena. Everyone can come to this conclusion by making logical deductions based on the universality of the laws. Another necessary step is then to understand that the visible levels are *not* more important than the invisible levels because there are *no* privileged places in the natural order—the laws deal with everything with perfect equality, without ever making an exception.

In the field of physics, this idea that laws must always apply in the same way is not only called the universality of the laws of nature, but also the *principle of relativity.* Everyone has already heard of relativity, especially because of the success of a certain Albert Einstein, who made it the center of his theories.

A common conception of the principle of relativity is to say, "Everything is relative," but this idea is not the essence of relativity. This is because it is not true that everything is relative; some realities are absolute, which means they never change, regardless of the way you look at them. Again, these realities are the *laws of nature.* That is why if you ask a physicist to define the principle of relativity, he will not tell you, "Everything is relative," but rather, "The laws are the same for all frames of reference." Because, even though a phenomenon may appear very different depending on the frame of reference, that is to say, according to the point of view, the laws that govern it remain exactly the same.

Because of this, a more complete definition of the principle of relativity would be, "The laws are absolute; the rest is relative." In fact, one could just as easily say "principle of universality," instead of "principle of relativity." These expressions are only different ways of naming the most fundamental principle of science.

At the heart of physics, there is the universality of the laws, but also the notions of frames of reference, points of view, or contexts, which are very important. This is because the description of a phenomenon is valid only within a certain context. Once again, it is a simple and natural idea that everyone already knows intuitively.

For example, if you stand in front of someone who throws you a ball, you will describe the movement of the ball by saying that it has come directly to you, while the person who threw it at you will say that it moved away from him. If there are outside observers, some will say that the ball has moved from left to right while others, on the other side, will say that it has moved from right to left. If some observers are moving, they will describe the speed of the ball in a different way than those who are motionless, and so on. Different observers may, therefore, describe the same phenomenon differently, each from its own point of view, and these various descriptions are equally valid. This relativity of motion means that to describe a motion, we must first define an observer or a frame of reference, from which this phenomenon will be described, and therefore, that a description is valid only within a certain context.

In his theory of special relativity, Einstein pushed this idea much further, demonstrating that even the notions of space and time may change among observers because the laws dictate that the speed of light must always stay the same across all frames of reference. In other words, according to Einstein's relativity, space and time are *relative* concepts and not absolute concepts because only the laws of nature are absolute.

This is a fascinating subject, but it would be too long to cover it in detail here. What must be remembered is that this is another proof of the power of the principle of universality, or of relativity. The essence of Einstein's genius was to understand the importance of this principle, and it is for this reason that his two great theories, special relativity and general relativity, are named after it.

From Galileo in the 17th century to Einstein and his successors, the principle of relativity has always been one of the primary pillars of physics. One of the most important lessons of this principle is that all frames of reference are equivalent, as long as they respect the laws of nature. All points of view are equal because no point in the universe is more important than another for the laws of nature. Relativity is sometimes interpreted, in a very superficial way, as proof that there are no absolute truths, only various opinions; whereas, in reality, this principle tells us that there are indeed absolute truths, *the laws of nature,* and that it is these laws that must be placed at the center of our worldview, not appearances.

So, what is the relationship between the notion of frames of reference and the visible and the invisible? These notions have a lot in common because the visible and the invisible are just frames of reference, points of view, contexts...and nothing else! The visible is only a certain point of view that we have on reality, a frame of reference which, like all frames of reference, *does not have any particular importance.*

Believing that the visible is all that matters and that the invisible is insignificant is in contradiction with the universality of the laws since it makes the visible a privileged frame of reference, a point of view more important than the others, which is forbidden by the principle of relativity. In other words, to believe that the visible is particularly important, one must throw into the trash the most fundamental principle of science, only to satisfy appearances!

Materialism is a belief system based on inconsistencies, logical flaws, which, sooner or later, will bring about the fall of this false philosophy. One of the greatest faults of materialism is *the belief that the visible is more important than the invisible,* a belief that contradicts the fundamental lessons of science.

From a symbolic point of view, putting the visible at the center of our conception of the world is the same as placing the Earth at the center of the universe. In both cases, it produces a *distorted* vision of reality, which gives more importance to certain deceptive appearances than to the laws of nature. Historians of science call "geocentrism" the belief system in which the Earth is the center of the universe, and in the case of materialism, one could just as easily call it "visiocentrism." Namely, a conception of the world that considers the visible, what is directly measurable, as the center of reality.

On the contrary, this view, which gives importance only to what is directly measurable, leads to a disconnection of reality, since it leads us to believe that what cannot be measured is less real than what is accessible to our instruments. A conception of the world that is very pretentious, since, for this position to be tenable, we must believe that reality conforms itself to the limits of our instruments, which is completely absurd!

The invisible part of reality, the one that our instruments cannot measure directly, will always be more important than its visible part. That is why it is not upon the visible on which we must construct our theories, but on the laws, since the laws are immutable while the visible is a category that changes according to the era. Many previously invisible realities once considered "nonexistent," are today considered facts, without which it is impossible to understand the world. It is the same today: The invisible still has many answers to give us!

At the beginning of this section, we talked about the belief that we cannot know anything about the invisible, a belief that is widespread among materialists. We have seen that this prejudice is false, since, by making deductions based on the universality of the laws, we can learn a lot about the invisible.

First, we have seen that the invisible is necessarily made up of levels, just like the visible. Second, those levels are just as diversified as the visible levels since the same laws that give rise to an infinite variety of phenomena in the visible levels must also apply. Third, the visible does not have a special place in the natural order, even though, from our point of view, it seems to be the case. The visible is only a certain point of view of reality, and the invisible is the combination of all the levels to which we do not have access.

We will now see one last fact that the universality of the laws allows us to understand the invisible, the fact that consciousness and life necessarily come from the invisible side of reality.

The laws of nature must apply universally. That is the essence of the principle of universality. This means that if a law applies to a category of phenomena, it must apply to all phenomena included in this domain, always in the same way, without exception, regardless of the context, anywhere in the universe, as much in the past, the present, or the future...

In other words, it is the laws that *define* the categories, and if a phenomenon belongs to a certain category, it must always obey the laws that define this domain, *without any exception*. For this reason, it is very important to classify the phenomena in the right categories since it tells us which laws should apply to these elements.

Within the universalist approach, the classification of elements is always clear. Nothing is more important, since, as we have seen in the preceding chapters, correctly classifying concepts that were badly placed before can allow us to take giant steps forward in our understanding of the world and can even solve certain "great mysteries" that were artificially created by our bad theories.

To solve mysteries in this way is what the universalist approach allows us to do, by correctly classifying these concepts that are the brain and the first life forms, a classification that show that the explanations of consciousness and life must necessarily be on the invisible side of reality. This is by following certain reasonings based on the principle of universality, which forbids exceptions to the laws of nature. Here are these reasonings, in the form of premises and conclusions:

1: Every material object is unconscious; it is a law of nature.

2: The brain is a material object.

Conclusion 1: Therefore, the brain is unconscious.

1: If the explanation of consciousness is not in the visible, it implies that the explanation of consciousness is in the invisible.

2: The explanation of consciousness is not in the visible because the brain is unconscious, as are all material objects.

Conclusion 2: Therefore, the explanation of consciousness is in the invisible.

1: Every life form is the reproduction of a previous life; it is a law of nature.

2: The first forms of material life, or visible life, are life forms.

Conclusion 3: Therefore, the first forms of material life, or visible life, are reproductions of a previous life.

1: If the explanation of the origin of life is not in the visible, it implies that the explanation of the origin of life is in the invisible.

2: The explanation of the origin of life is not in the visible because the first forms of material life, or visible life, are reproductions of a previous life, as are all life forms.

Conclusion 4: Therefore, the explanation of the origin of life is in the invisible.

Final Conclusion: Therefore, the universality of the laws of nature necessarily implies that the explanations of consciousness and the origin of life are in the invisible, since any other conclusion requires exceptions to the laws.

In the eyes of many, it will seem that these reasonings go too far, that they go beyond what science allows us to affirm. But, in this book, everyone will be able to understand that there is no better way to create a worldview that is 100% consistent with science. These conclusions are consistent with science because there is no more solid basis for building scientific theories than the principle of universality!

Universalistic reasonings are simple to follow, and, of course, materialists can reply by trying to convince us that those subjects are not so simple. In particular, they can try to convince us that it is not true that all material objects are unconscious and all life forms are reproductions. In other words, they can try to make us believe that these rules are not true laws of nature, and therefore, exceptions are allowed. For this, they are ready to use all kinds of intellectual contortions.

This is because the idea that all material objects are unconscious and that all life forms are reproductions of a previous life are not laws invented by non-materialists to protect their beliefs. On the contrary, these are laws that everyone, *even the materialists,* continually apply! Indeed, even materialists consider that all material objects are unconscious, making an exception only for the brain. Likewise, they regard all life forms as reproductions, making an exception only for what they think are the first life forms. So, even materialists, in their normal reasoning, regard these rules as laws of nature, and they are inconsistent with these rules only when it is necessary to protect their beliefs. In these cases, they follow abnormal reasonings, just like religious people who believe in miracles.

To see it, one just needs to ask the materialists to explain how such exceptions to natural laws are possible. That is when the show of intellectual illusionism begins, since this is impossible to explain without contradicting laws that everyone normally considers as universal! In their explanations, materialists can only try to save appearances, relying on all sorts of tricks to hide the fact that their reasonings are never truly coherent.

Materialist scientists share the belief that consciousness and life come from matter, but when they try to understand how it is possible, all they generate are endless debates that lead them nowhere, and this is why these topics are still considered the greatest mysteries of science. They are lost in a labyrinth created by their false beliefs. From the universalist point of view, the origin of this confusion is easy to understand: *It comes from the fact that materialists believe that exceptions to the laws of nature are possible!*

Materialist theorists and philosophers can be criticized, with their lame attempts to explain the exceptions they allow in their theories, but at least these people try to give some rational frame to their beliefs. This is not the case for the vast majority of materialists, who simply believe that consciousness comes from the brain, and that life comes from matter, only because, from a certain point of view, this interpretation is consistent with appearances. These people never make an effort to explain their beliefs rationally. *They only believe!* They believe in the miracles of the brain and of matter, just as other people believe in the miracles of their religion. Materialistic thinking is just a form of magical thinking, in which one grants imaginary powers to matter rather than to a sort of deity!

It is *only* to respect the laws of nature that the universalist approach proposes invisible solutions. Materialists often claim that, for non-materialistic solutions to be true, it would be necessary to overturn what science knows about reality, but it is easy to see how false this prejudice is. Again, all that is overturned are certain materialistic beliefs, which many intellectual illusionists present to us as "scientific facts." No discovery of science is overturned by the universalist approach. On the contrary, its goal is to agree with the most important laws of science!

Science agrees very well with the idea that most of reality is invisible, and the law of selection allows us to explain naturally how this is possible. Seeing the brain as an unconscious object like any other allows us to solve one of the greatest mysteries of science since we no longer need to search by which mechanism this object can generate consciousness. The universalist approach also allows us to always respect the law that tells us that every life form is a reproduction, a law that is certainly one of the most well-tested of all science! These are just some of the reasons this approach is an excellent solution. It is not necessary to abandon anything of science, only to build on its strongest pillars!

The theory presented in this book is not special; it is constructed as any scientific theory must be. This is because we start from well-tested laws, on which we gradually build a representation of the world that is consistent with these laws. The universalist approach has an easy-to-understand basis, but that does not mean that this theory is simplistic—it has a simple basis because it is so for all scientific theories! One only needs to study the theories that have become well established throughout history to understand that it is always so. At the base of these theories, we continually find a handful of elementary rules, which served as a starting point for these models.

For example, the Newtonian theory of gravity is based on the idea that all objects behave in the same way concerning gravity, regardless of whether it is an apple or the moon, a simple rule, which was revolutionary at the time. Modern chemistry began when researchers realized that all elements of nature are composed of the same sub-elements and that what distinguishes one element from another is only the number of protons contained in its core, a much simpler conception of the elements than the one previously maintained. In biology, the concept of natural selection, essential for understanding the evolution of species, is also elementary...

The laws underlying scientific theories are always simple because the laws of nature are always simple! It is only the *logical consequences* that accompany these rules that can be complex, and sometimes very astonishing.

For example, Einstein's relativity tells us that space and time can change from one frame of reference to another, which is very surprising since we are used to believing that these notions are immutable. However, this logical consequence of relativity does not mean that this theory has a complex basis since it is a consequence of a simple principle: the fact that the speed of light must be the same for all frames of reference. Similarly, quantum physics has a reputation for being complex and incomprehensible. Yet, like all theories, it also has a simple basis. For example, its laws tell us that energy can never be exactly zero. Another simple idea, which nevertheless, has astonishing logical consequences since it means that the void does not really exist! Indeed, the void cannot exist, since there must always be the presence of a certain amount of energy, according to the laws of quantum physics. That is why, in the theories of modern physics, emptiness is rather seen as a state of minimal energy, and not as "the absence of everything."

Science is filled with amazing facts, but that does not mean that the laws at the base of science are complex and incomprehensible, only that some of our prejudices must be reconsidered! It is the same with the idea that consciousness and life have their source on the invisible side of reality. It is only a logical consequence of the laws of nature, like the two preceding laws: Every material object is unconscious, and every life form is a reproduction.

The conclusion that consciousness and life come from the invisible part of nature can be amazing for some people, but it is certainly not a stranger concept than others that are commonly accepted by scientists of our time. As another example, the formulas of general relativity tell us that it is theoretically possible to compress the Earth until its entire mass is contained in a space the size of an egg. Can we imagine a concept harder to believe? Yet, it is an idea considered banal by the physicists of our time, many of whom believe that the existence of invisible life forms is a far-fetched concept! People who do not believe that an invisible life is possible only emit a judgment based on their prejudices, and not on reality, which allows all kinds of much more extraordinary phenomena. Life is present everywhere on our planet. What is so strange about the idea that it is also present in the invisible substances that surround us? What criteria do materialists rely upon to say that it is impossible, except the fact that it contradicts their preconceived idea of life?

A logical consequence can only surprise us when it contradicts our prejudices, and it is the same for the existence of invisible life forms. In reality, this concept is natural since it is a logical consequence of the laws of nature.

6.4 IN SUMMARY

Using invisible solutions to answer scientific questions is allowed, as long as we have good reasons to do so!

As we have seen in this chapter, there are many good reasons to use the invisible to solve the mysteries of consciousness and the origin of life: first, because the visible solutions are unsatisfactory, otherwise these subjects would not be considered the greatest mysteries of science; second, because throughout the history of science, many invisible solutions have been successful; third, because it is allowed to postulate the existence of invisible substances or forces when necessary to answer certain questions, as we have not yet found a limit to the number of particles and forces that can exist; fourth, because the law of selection allows us to explain naturally why these invisible realities can escape our instruments; fifth, because a lot of indirect evidence tells us that most of reality is invisible; and sixth, because resorting to invisible solutions is necessary to respect the universality of the laws of nature.

Universalism is based on many *reasons,* which is why it is a *rational* position. Traditionally, people have the habit of thinking that invisible solutions to the enigmas of life are in the realm of faith, while visible solutions are considered to be in the realm of science. But this categorization is false since it is possible, only by using logical deductions based on the laws of nature, to come to the conclusion that the solutions to the mysteries of consciousness and of life are necessarily invisible. No act of faith is required for this; it is only a question of logic! To see more clearly, all one needs to do is to organize some key concepts correctly. To help us do this work, here are the primary logical relations and conceptual unifications we saw in this chapter:

The invisible is not supernatural.

The invisible is what lies outside the limits of our instruments.

The visible is not the center of reality.

The visible is what lies within the limits of our instruments.

The invisible is larger than the visible.

The universality of the laws encompasses the visible and the invisible.

The visible and the invisible are levels of reality.

The visible and the invisible are diversified domains.

The explanation of consciousness is in the invisible domain.

The explanation of the origin of life is in the invisible domain.

Whoever neglects the importance of the invisible *cannot* construct a good representation of reality. By placing the visible at the center of their theories, instead of the laws of nature, materialists have created a distorted worldview filled with artificial mysteries. They like to claim that they are "realists," but their conceptions only disconnect them from reality, which is essentially invisible. The universalist approach tries to restore order, returning the invisible to the place it belongs in the natural order: The invisible is the most important part of nature, and this domain is necessary to understand consciousness and life.

7. THE IMPORTANCE OF ENERGY

The essence of reality is energy, not matter.

So far, we have seen two of the fundamental ideas that are behind the universalist approach: the universality of the laws of nature and the importance of the invisible. These are two indispensable elements to answer the enigmas of consciousness and life, and we will now see another one, which is energy.

In this chapter, we will explore the real importance of energy, which will allow us to understand better why it must be placed at the heart of our theories seeking to explain consciousness and the origin of life, just as energy is at the heart of the greatest theories of science. Let us continue our reflection, seeking to go beyond certain misconceptions based on appearances.

When studying physics, an assertion that one sometimes sees is that matter is essentially empty. But what does this really mean? To represent this emptiness, we can visualize what an atom would look like if we could observe it very closely. An atom is composed of a nucleus, which concentrates almost all of its mass, around which there is a layer of electrons. But the electrons are not close to the nucleus; between them, there is a space that is vast, proportionally to the dimensions of the atom. Indeed, if we could magnify the nucleus of an atom to give it the size of a pea, the electronic layer would be the size of a football stadium! This means that if we could see the matter of a solid object closely, the atomic nuclei would appear to us like peas separated by hundreds of meters of distance, a huge empty space that the tiny electrons inhabit like so many little flies.

The mathematical formulas that describe this domain are difficult to translate into images, so we must not take these metaphors literally, but rather see them as representations that put forward certain aspects. Peas and football stadiums are not very precise units of measurement, but this is enough to give us an idea of the relationship between the space occupied by particles of matter and the empty space. So, if we want to visualize what matter looks like inside a solid object, we should not imagine the atoms crammed onto each other like balls in a box, but rather imagine that they vibrate next to each other, being separated by a space that, relative to the nuclei of the atoms, is immense.

Another way to illustrate how empty is the world around us is to try to imagine what would happen if we tried to compress an object to its highest compaction level. For example, if we take an empty cardboard box, we know that if we compress it to remove the empty space it contains, we can make it occupy much less space. Suppose we are really zealous and we want to compress our box in the smallest possible space to store it, and that, for this purpose, we decide to ask a physicist what is the smallest possible space that our box can occupy when compressed to the maximum. Again, the answer may surprise us because it is theoretically possible to compress our box into a microscopic space. By subjecting it to enough pressure, we could even compress it into the space normally occupied by a few atoms of matter! Conclusion: By compressing them to the maximum, we could put all the boxes of the world in a box of matches.

It may be hard to believe, but there are really objects in the universe that have this extreme density. These form at the death of stars, when the processes of nuclear fusion normally occurring in their core are exhausted, and can no longer generate enough energy to counterbalance the incredible pressure coming from the mass of these celestial bodies. These stars then collapse on themselves, reaching an extreme level of compaction. The densest objects that can be formed this way are the black holes, celestial bodies that emit no light, and have an unimaginable density. The density of the black holes is so high that if we could compress the planet Earth to this level of compaction, the whole of its mass could be contained in a sphere with a radius of nine millimeters—smaller than an egg!

It may seem incredible, but that is what the formulas of physics say. These are not marginal theories, but concepts that have been known to physicists for a long time. This radius of nine millimeters in which we could compress the Earth is called the "Schwarzschild radius," which represents the density at which an object becomes a black hole, the densest object in the universe.

If the whole mass of the Earth can be contained in an egg, it means the space it currently occupies is essentially empty...or, to be more precise, that this space is occupied by something other than matter. What occupies this void between the particles of matter? If atoms are separated by so much space, what keeps them from falling on one another? Why do objects seem solid to us, if they contain so much emptiness?

To answer these questions, it is necessary to remember that the particles of matter are not alone; they are accompanied by another category of particles that transmit the interactions, mediating particles that are the vehicles of the forces of nature.

The dimension that the Earth would have, if it had the density of a black hole.

If atoms behave as they do, if they vibrate next to each other, leaving a lot of space between them, it is because it is imposed on them by the forces that animate and structure matter. The presence of these interactions means that the void between the particles of matter is not really empty. This space is occupied by *force fields,* and it is this network of interaction that dictates to matter how it must behave. What maintains the cohesion of the objects is not the matter itself, but the forces conveyed by the mediating particles, particles that are of the same nature as light. We can also visualize this exchange of particles as a wave transmission process, within which the particles of matter act as antennas or relay centers.

What keeps the cohesion of matter are these waves and these mediating particles, and not the particles of matter themselves, which are inert and only react passively to the action of the forces. If the objects seem solid to us, it is only because the forces that exist *between* the particles of matter cause them to repel or attract each other, processes that give a structure to the objects. If we remove these forces, matter collapses on itself because it is essentially inert.

These mediating particles that flow between the particles of matter can also be called particles of force or particles of *energy,* and because of that, we can say that the universe is composed of two primary ingredients: energy and matter.

To say that energy is an ingredient, a substance made up of particles, is another statement that can be surprising. In physics, energy is considered a unit that measures the ability to produce a displacement, as the meter is used to measure lengths, or the second is used to measure time. Energy is then defined as a measure of "work capacity."

Alternatively, in quantum physics, the field of physics that describes the behavior of nature on a very small scale, energy is also considered in a more elaborated way because the classical approach to energy had proven insufficient. To explain certain phenomena, the solution that physicists have found is to consider that energy exchanges, in the domain of the extremely small, are in the form of *particles,* in packets called quanta. This realization is the basis of quantum physics, and why this branch of physics is so named.

The energy packets, the quanta, have an energy that is always a multiple of a quantity called "Planck's constant." This quantity is insignificant on our scale, but at the particle scale, it has a huge effect because the fact that all the events that occur in the world of particles must possess an energy that is a multiple of this constant draws a precise limit between the phenomena that are possible and those that are impossible.

The best-known of these quanta of energy is the photon, the particle of light, hence another crucial concept: The notions of energy and light are very close to each other, an intimate relationship that is crystallized in the formula "$E=mc^2$." Which tells us that to know the amount of energy that an object contains, it is necessary to multiply its mass by the square of the speed of *light*.

The field of quantum physics has another similar equation, which is written "$E=hf$." In this equation, "h" is the Planck constant, and "f" is the frequency. This formula tells us that the amount of energy contained in a mediating particle is related

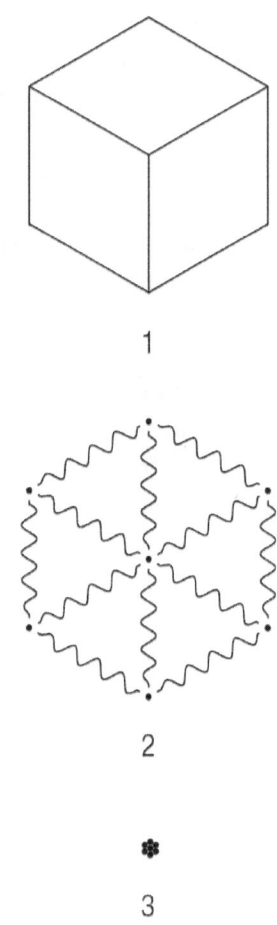

The space occupied by the objects seems to be filled by matter (1), but in reality this space is essentially occupied by the energy that is present between the particles of matter (2). If this energy is removed, matter collapses on itself until it occupies almost no space (3).

to its frequency, a relationship that allows us to quantify how much energy it carries. In short, one of the revolutions of quantum physics has been to give a *substance* to energy, a quantifiable substance of the same nature as light, and no longer to consider it merely as an abstract unit of measure.

Very summarily, we could sum up quantum physics in these terms: *Everything is made of particles.* Matter is made of particles, and energy is made of particles. In a way, the particles can be considered the "pixels" of reality since, just as the images on a screen are composed of pixels, all that exists is composed of particles.

One could also summarize quantum physics in these terms: *Everything is made of waves.* Matter is made of waves; energy is made of waves; there is nothing that cannot be considered as a wave phenomenon.

At first glance, these may seem like contradictory affirmations. How can a phenomenon be a particle and a wave at the same time? Particles are clearly defined phenomena, whereas waves are spread out in space. Normally, we use one or the other of these concepts to describe the phenomena around us; to our eyes, either they are clearly defined as a particle, or they are diffused like a wave. But, in the quantum domain, it is not possible to completely understand a phenomenon without needing to use both of these concepts because to describe the phenomena only in terms of particles, or only in terms of waves, gives us only a partial image that is not enough to explain all the observations. As a result, we must always consider that quantum phenomena have both particle and wave characteristics. This is the "wave-particle duality," another important conceptual unification of modern physics.

In reality, there is no opposition between these two visions since they are *complementary.* Just as an object may seem different depending on whether we look at its face or its profile, the quantum objects can appear to us differently depending on how we observe them. That is why, in physics experiments, the same phenomenon may appear to us to be composed of particles if the measurements are taken in a certain way, and as a wave phenomenon if the measurements are taken in another way. This is also why, when we talk about energy transmissions, we sometimes consider them a transmission of particles, and sometimes a transmission of waves, since both interpretations are correct. In reality, to say that a phenomenon is composed of particles or waves is dependent on our point of view! Again, those are relative notions.

What is most important to remember is that the particles of matter never interact directly with each other; they interact only by the exchange of particles of energy, of light pulses, that we can also visualize as a wave exchange. We can also see this circulation of quanta of energy as the exchange of bits of information that occurs between computers because the quanta also carry information and the effects they produce while circulating between the atoms and the molecules are not fuzzy, but very precise. From this point of view, we can see each atom and molecule as tiny antennas that constantly receive and send information through photons of light. A photon of a precise energy that reaches an atom or a molecule changes its configuration precisely, which in turn, modifies the behavior of this element precisely. In this way, the behavior of matter is literally programed by the quanta of energy circulating in it.

This behavior is governed by a great law of nature, the law of interactions, which can be summarized as follows: *Every interaction is an exchange of energies that carry information.* The universe is not an incomprehensible chaos but a structure rigorously ordered by the laws of nature, and what maintains this order is the energy that circulates everywhere, conveying information.

Energy is the *essence* of reality, whereas matter is a secondary element that depends on energy for its existence, as a shadow depends on its source. It is energy that structures matter, and it is energy that dictates to matter how it must behave. The only intrinsic property of matter is *inertia,* passivity, resistance to change. All other qualities, such as colors, sounds, flavors, warmth, shapes, movements, are the product of an *interaction* between matter and energy. If this energy is removed, matter collapses on itself until it occupies almost no space, as shown by the example of black holes.

Matter is inertia; everything else comes from energy, from light. Therefore, it is quite logical that it is also the case for consciousness and life: *They come from energy, not from matter.*

When we try to explain the origin of consciousness and of life by relying on matter, we only encounter "mysteries" because matter contains *nothing* that can give us an explanation. It is only an inert element, reacting passively to the impulses it receives. Alternatively, if we rely on energy, natural solutions are available to us. As we will see, there are many good reasons to place energy at the center of our conception of consciousness and life, just as we did with the invisible in the previous chapter.

7.1 UNDER PRESSURE

The universe is essentially composed of energy. We can use the equations "$E=mc^2$" or "$E=hf$" to define energy, but one could simply say "$E=R$," or "Energy = Reality." Energy is the fundamental element of nature. Instead of the laws of nature, we might as well speak of the laws of energy, or the laws of motion, these expressions are synonymous.

Without energy, there is nothing. Even matter owes its existence to energy, since it cannot be formed if there is not at first the presence of a certain concentration of energy. Even if from a certain point of view we consider that energy and matter exist separately, from another point of view matter is only an effect of the activity of energy and has no independent existence.

Matter is only a product of energy. It is this process that occurs within particle colliders, where the collision of particles at speeds close to the speed of light releases a large amount of energy, which quickly gives rise to myriads of particles of matter. It is also this process that occurred at the origin of the universe, at the time of the Big Bang, where everything was initially only energy, an energy that has generated various forms of matter with the cooling of the universe.

All the structures that surround us, from particles to galaxies, are the product of an equilibrium between the forces of nature, forces acting through the radiation that fill the universe. These energies, these powers, these force fields, are present everywhere and are of an inconceivable intensity. Nature may seem peaceful at one level, but when you look closely at the game going on between the forces of nature, the picture can be quite different. If we usually do not perceive the presence of the unimaginable pressure to which the forces of nature subject everything, it is because the effects of these forces tend to cancel each other out to create states of equilibrium.

We can illustrate these equilibrium states by a well-known example of pressure: the atmospheric pressure. We live at the bottom of an ocean of air, and the pressure that the atmosphere exerts on us is very high, about one kilogram per square centimeter. Without noticing, we all have several hundred kilos pushing on our shoulders! How is it that we do not feel the enormous pressure of these kilos of air? Because the pressure of the gases and liquids inside our bodies is the *same* as that of the atmosphere.

This is mathematical: If a force of one kilo is exerted in one direction, and it encounters a force of one kilo exerted in the opposite direction, the result is equal to *zero,* which creates the illusion that there is no force at work. The result is the same if we oppose a force of one thousand kilos to another thousand kilos, or if we oppose forces of millions of kilos... It is always the same zero, because mathematics does not discriminate!

When two identical forces oppose each other, a state of equilibrium is created, which gives us the illusion that there is no force at work. On the other hand, as soon as one of the two opposing forces is removed, the other force can then manifest all its power. This is done, for example, with suction cups. When we remove the air under a suction cup, nothing is going against the pressure of the atmosphere anymore, and it is this pressure that forces the suction cup to adhere to a surface. But suction cups are not a very effective way to create a vacuum. With a pump, one can achieve much more impressive results, as shown by the most famous experiment showing the power of atmospheric pressure, the experiment of the "Magdeburg hemispheres," which took place for the first time in the 17th century.

In this experiment, two metal hemispheres were joined and sealed with grease; then, using a pump, all the air inside was removed. The result was a sphere the two halves of which were held together *only* thanks to the atmospheric pressure. One might think that the pressure of the air is not enough to make this sphere resistant, but it would be a mistake because two teams of horses, pulling in opposite directions, did not manage to separate it! This is not surprising since the pressure that held these two hemispheres together was about two tons... The most impressive in this story is not the fact that horses fail to separate two hemispheres held together only by the pressure of the air around them, but the fact that we are constantly subjected to this pressure without noticing it! Because it means that we are *ourselves* filled with a pressure comparable to that of the atmosphere.

The pressure that the atmosphere exerts on us is high, but it is still nothing. The situation is much worse at the bottom of the oceans, where the pressure can exceed a thousand kilos per square centimeter. This pressure that exists in the deepest parts of the ocean is the reason the exploration of these environments is extremely difficult, because we have to build submarines specifically designed to withstand these gigantic pressures. However, while humans venture into these depths, fearing to be crushed if their submarine fails, they meet in these places fishes that go around peacefully, completely unaware of the pressure of several tons to which they are subjected. The reason fishes on the bottom of the oceans do not feel this pressure is the same reason we do not feel the pressure of the atmosphere: because they are themselves filled with water, which exerts inside their body the same pressure they are subjected to, which cancels out the effect of the pressure of the ocean. Just as we are filled with a pressure comparable to that of the atmosphere, which allows us not to be crushed by it, these fish are filled with a pressure of several tons, and that is the reason they are not crushed by the ocean. If it was possible to bring these fishes back up to the surface in a fraction of a second, so that all the pressure they contain was suddenly released, they would explode like dynamite!

Just as we live in an ocean of air and fishes live in an ocean of water, the whole universe is bathed in an ocean of energy. This energy permeates everything. In the form of light, it fills the space between the stars and is present between each particle of matter, where it acts as the vehicle of all interactions. Like the atmosphere and the oceans, which exert gigantic pressure on their inhabitants without them being aware of it, the omnipresent energy maintains levels of pressure and tension that are inconceivably intense. Again, because of the balance of the forces, we do not usually feel this intensity that permeates everything. On the other hand, one only has to break this equilibrium to see this energy reveal itself in all its power.

It is very easy to observe the energy that is present everywhere. For that, we only have to ignite an object. By doing this, we cause a chain reaction that releases the energy that was previously used to bind the molecules of this object together. This energy that comes to us in the form of heat was always present in the object and the oxygen from air. Only, it was previously kept under control by the balance of the forces of nature. In this case, it was the particles of matter that acted like a container that prevented the energy from escaping.

All the elements, objects, shapes and structures that surround us are born of a state of equilibrium between the forces of nature. This balance can be very stable, as in a stone; or very precarious, as in an explosive. In the case of a precarious balance, a simple spark can be enough to release the energy with a lethal force, like a step in the snow can be enough to trigger an avalanche.

To ignite an object is nothing more than to break the balance of forces that previously held the energy inside. The fact that there is almost nothing left of an object after it has been burned shows us that, in reality, this object contained very little matter; it was essentially energy. Just as the space occupied by a balloon is, in reality, occupied by air particles, the space occupied by an object is essentially occupied by energy particles, waves, force fields. The smoke that flies away and the ashes that remain is all the matter an object contains, the rest is only light.

This release of energy in the form of flames may seem trivial (except when it is our house that is burning), but it is just a tiny sample of the energy that permeates everything. By breaking the bonds between molecules and atoms, we can release a lot of this energy that was between the particles of matter, but if we go further, attacking the nuclei of atoms, the result is even more spectacular.

The nuclei of the atoms are composed of protons and neutrons, particles that are also connected to each other by energy. On the other hand, this energy is by far superior to the energy that serves as an intermediary between the atoms and the molecules. The most famous and unfortunate expressions of this energy are nuclear weapons, which succeed in breaking, by a chain reaction, the nuclei of atoms to liberate this energy in a destructive way. In these cases, the energy contained in a handful of material may be enough to level a city! These bombs show us, with painful clarity, the colossal amount of energy that is present everywhere; they are terrible experimental proof of the equation "$E=mc^2$."

Nuclear weapons give us a bad image of nuclear energy because it is not only destructive. On the contrary, it is thanks to nuclear energy that there is life on Earth because it is the nuclear reactions that occur in the heart of the Sun that feed our planet with energy. Alternatively, it is by fusing nuclei of atoms and not by breaking them, that the Sun produces energy.

Simplifying, we can consider the Sun as an immense ocean of hydrogen, in which there is enormous pressure, as is the case in the oceans of water and air that are on Earth. This ocean of hydrogen is millions of times deeper than those found on Earth, and the pressure, in its depths, is millions of times stronger, which forces the protons, which normally repel, to merge to form new elements. The cores of the stars are the only places where there is naturally enough pressure to form atomic nuclei. This explains why almost all the elements have been formed in the stars.

Every particle of our body, every atom we breathe, was formed in the heart of a star at a very distant time. When we play the sorcerer's apprentices, releasing the energy of atomic nuclei during nuclear explosions, it is literally the energy of the stars that we release, a tiny sample of the intensity that exists within these celestial bodies.

Examples of the omnipresence and the power of energy abound, but none exceed in intensity the supernovas, the explosions of stars. Although they can live for billions of years, stars are not eternal. There is inevitably a point in the life of a star where it exhausts its fuel. That is to say, it has no more particles to merge to release energy. The slowing of the nuclear fusion process results in the radiation pressure from the star's core no longer counterbalancing the pressure exerted by the ocean of matter in the star. There is then a rupture in the balance of forces that existed before, and, as was the case in the previous examples, a sequence of spectacular events results from this rupture.

Under the pressure of its own mass, the star is compressed more and more, and the unimaginable pressure that exists in this core becomes even more unimaginable. The sequence of events that occurs at the end of a star's life varies greatly depending on the initial mass of the star in question. In the case of stars much more massive than the Sun, the sequence of events resulting from this compression can produce a gigantic explosion that blows out the outer layers of the star by spreading them in the interstellar medium. The term "gigantic explosion" is a euphemism since this explosion releases more energy than the radiation of the stars of an entire galaxy! By exploding in this way, the star releases all the elements it has formed during its life, elements that will be used to form new celestial bodies. So, we must not see the death of these stars as a catastrophe but as a ripe fruit that generously spills its seed. Star explosions are both the most terrifying and most magnificent events in the universe.

From the nuclei of atoms to the explosions of stars, the amount of energy that is everywhere present is absolutely inconceivable. If we look at the universe from the point of view of matter, we can say that it is essentially empty, but if we look at the universe from the point of view of energy, the portrait is the opposite: The universe is filled with energy, with light. Nothing is empty, everything is filled with radiation, and it is the activity of these force fields that determines the structure and functioning of the universe.

7.2 BOSONS AND FERMIONS

To understand better the fundamental difference between the energy particles and those of matter, we will now look at how physicists describe the behavior of particles. To do this, we will delve into two important concepts of particle physics: bosons and fermions.

These concepts are used to describe the two types of particle behavior that exist: A particle has either a bosonic behavior or a fermionic behavior. Only, instead of the word "behavior," physicists use the word "statistics."

To be more precise, the bosons are the particles that obey the Bose-Einstein statistic, while the fermions are those that obey the Fermi-Dirac statistic.

The concepts of bosons and fermions have strange names, and because of that, they may seem more mysterious than they actually are, but these names only come from the people who participated in their discovery, that is, Satyendra Nath Bose in the case of the boson, and Enrico Fermi in the case of the fermion. These are just names chosen to honor their discoverers, names that tell us nothing about the characteristics of these particles.

These names could have been quite different since, in science, names are nothing sacred; they are conventions that are always more or less arbitrary. For example, the word "atom" comes from the Greek language and it is a word that means "indivisible," a name that is false since we now know that atoms are composed of sub-particles. Yet, despite this discovery, scientists have decided to keep this name because it is a habit and because it would be too complicated to agree on a better name.

Similarly, when physicists discovered the proton, one of the particles that make up the nucleus of atoms, they chose this name taking from the root word "proto," which means "fundamental." This is because, at the time, they really thought they had discovered the fundamental particle of matter. Again, this belief was reversed when it was discovered that the protons themselves were composed of sub-particles, the quarks, a name that, for its part, is completely arbitrary... It is the same for the terms bosons and fermions, they are only names established by convention within a certain historical context, and that could have been quite different.

In the realm of elementary particles, fermions are the particles of matter, while bosons are the particles of energy or force. Except for the Higgs boson, physicists name them mediator bosons, or gauge bosons, since these are the elementary particles that convey the forces of nature. We can even consider all the mediator bosons as different kinds of light particles, since these bosons all have much in common with light. Physicists could very well have named the various mediating particles "light of type A," "light of type B," "light of type C," and so on; instead of naming them photons, gluons, gravitons and so on. The portrait would have been clearer since all these particles belong to the same family; they are cousins of the light particle. That is why, in this book, we will often use the word "light" to talk about bosons, just as we will use the word "matter" to talk about fermions, since these are the names that best help us to understand these notions intuitively.

So, what is the difference between these two kinds of elementary particles? The behavior of bosons and fermions differs on a crucial point, which is the following: *Identical bosons tend to unite, whereas, for identical fermions, it is the opposite; they cannot exist in the same place at the same time.* An analogy often used is to say that the bosons are "gregarious" particles, whereas the fermions are "solitary" particles. We can also see this as a form of innate attraction that exists between identical bosons, and a form of innate repulsion that exists between identical fermions.

To illustrate more clearly the difference between the behavior of bosons and that of fermions, it is necessary to deepen how physicists analyze the behavior of particles. For that, let us go back to quantum physics.

Because of the peculiarities of the quantum world, instead of describing the behavior of particles using the image of small balls that we commonly use, physicists describe phenomena using what they call the "wave function." This function represents all the possible states of a particle, in a diagram that has the structure and the functioning of a wave, and physicists use the wave function to extract the density of probability, which informs us about the most probable states of the studied system. Although the concept of wave function may seem very abstract, it is possible to visually represent a wave function since it can be graphically represented like any other wave. Here are some examples:

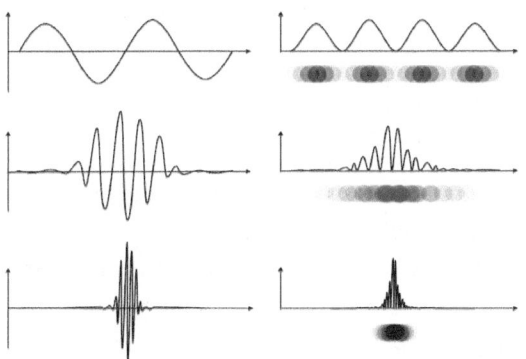

On the left, various simplified wave function diagrams; on the right, the density of probability associated with them. The vertical axis represents the probability level, and the horizontal axis represents the position. Under the diagrams showing the probability density, the opacity level of the spheres corresponds to the probability that the particle has this position.

As we see, in the case of the condensed wave function, the probabilities are very concentrated in one place. This means that the position of the particle described by this function is relatively well-defined. While in the case of more spread-out functions, the probabilities are diffused, and therefore, the position described by this function is fuzzy.

Based on this small sample of quantum physics, we can better understand the difference between the bosonic behavior, which applies to particles of energy or light, and the fermionic behavior, which applies to particles of matter.

We can use wave functions to describe a single particle, but we can also use it to describe a set of many particles. To do this, we can combine our particles to produce wave functions that will summarize our entire particle system. This is the power of this mathematical solution: *Wave functions combine like any other wave.*

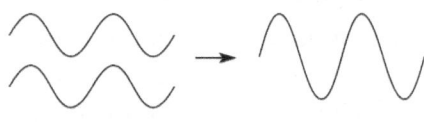

Constructive interference

Destructive interference

Intuitively, we all know what happens when we combine waves. For example, if we generate different waves on the surface of a container filled with water, we can observe how these waves will combine. Essentially, two things happen: when two peaks meet, they combine to temporarily produce an even higher peak, a phenomenon called "constructive interference." Whereas when a peak and a dip meet, they cancel each other out temporarily, which is "destructive interference."

By this simple game of addition and subtraction, the waves that we produce on the surface of the water will form a pattern, which is a summary of all the waves we have generated. The effects are similar in the quantum world. If we combine identical boson wave functions or those of identical fermions, these will interfere in a way comparable to water waves; however, the result will be different depending on the particle type. In the case of identical bosons, the resulting wave function will produce an *increase* in the probability of finding these bosons in the same place at the same time, whereas, in the case of identical fermions, it is the opposite: the probability of finding them in the same place at the same time *will always be zero*.

The fact that identical fermions cannot be at the same place at the same time is called the "exclusion principle," discovered by the physicist Wolfgang Pauli in the early 20th century; this principle is a pillar of quantum physics. The antithesis of this principle, the fact that identical bosons tend to unite, has no official name. If it were necessary to name this principle, it could be the "union principle," or the "bosonic union."

The fact that bosons do not resist one another, and instead, unite and work together explains why they are the vehicles of the forces of nature. Ultimately, what physicists call "forces" are only effects produced by the circulation of different types of bosons, waves or radiation that can attract and unite when they are in affinity, and what they call "force fields" are the areas where these influences are present.

The best-known example of a technical application of bosonic behavior is the laser. Lasers are able to use the natural tendency of light to vibrate in unison, to produce coherent radiation that can reach a very high intensity and concentrate a large amount of energy. Since there is theoretically no limit to the amount of energy that can be concentrated in one place, there is almost no limit to how much power a laser can reach. The idea, conveyed by science fiction, of laser weapons that can pierce anything, is not so crazy; the only thing that is needed is a device capable of producing and sustaining such intensity.

On the side of fermions, the principle of exclusion also plays a key role in many phenomena. In particular, this principle is essential to explain why electrons organize themselves in layers around atomic nuclei. Electrons are fermions, which means that two identical electrons cannot be in the same place around the nucleus. For this reason, we can see every possible state around the nucleus as a box, and when this box is already occupied by an electron, it forces the other electrons to go elsewhere. There are a limited number of these positions for each energy level, and when all of them are occupied at one level, this forces the other electrons to go to a higher level, it is through this process that electronic layers add up around atomic nuclei. Combined with the repulsion coming from the fact that electrons all have the same negative charge, the exclusion principle gives a form of rigidity to matter.

The behavior of bosons and fermions, despite the simple rules that define them, gives rise to a great variety of phenomena with nuances of all kinds. Moreover, it must be considered that most of these phenomena have not yet been discovered because they take place on the invisible side of nature.

7.3 ENERGY AND THE UNIVERSALITY OF THE LAWS OF NATURE

Energy is the largest domain of nature, and science is still far from understanding everything on this subject, which embraces the activity of light in all its known and unknown forms. Our ignorance is always much greater than our knowledge. However, using the principle of the universality of the laws of nature, it is still possible to grasp this domain in its broad outlines, as we did with the invisible in the previous chapter.

By applying the principle of universality to energy, we can come to several important conclusions. First, energy must comprise visible and invisible levels, just like matter. Second, the different levels of energy must be very diversified domains, just like the material domains. And also, the interactions between these different levels must be energy exchanges that carry information, just like the interactions within these different levels.

This worldview is summarized in the following diagram, which contains most of the concepts we have seen so far. This diagram is a symbolic representation that shows us only certain essential characteristics. In spite of its simplicity, this pictogram of reality contains everything we need to guide our thinking properly and allow us to answer many important questions, as we will see in the following chapters. This is because the answers to the great questions are not in the details, but in the great laws, the overview, the synthesis.

The invisible levels of nature are made of energy and matter, just like the visible

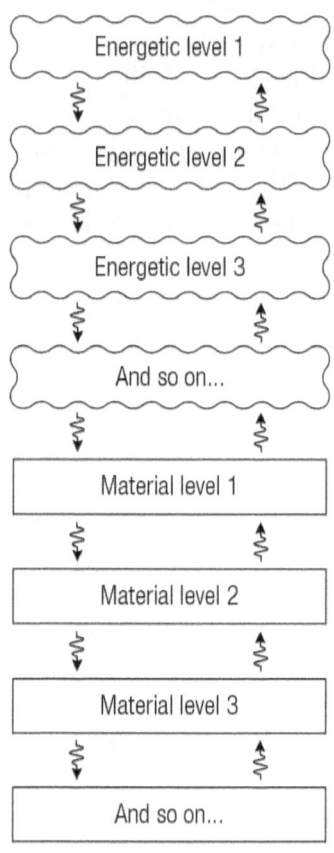

The universe consists of two main ingredients: energy and matter. Each of these categories is divided into many levels, formed by the law of selection; these levels interact with each other through various forms of energy, according to the law of interactions. Some of these domains are visible to us, but most are invisible.

levels. Everything that exists in nature is either made of energy, matter, or a mixture of both; there are no other options. Energy and matter are the two types of ingredients that form the universe, and the laws that apply to these fundamental types must also apply in the invisible, according to the principle of the universality of the laws of nature. Surely there exist in the invisible realms subgenres of energy and matter obeying laws that are peculiar to them. But this does not change the fact that these subgenres must always also respect the laws of their fundamental types because the laws of subgenres can only be added to the laws of the more fundamental types that encompass them. They can never contradict them.

It seems abstract presented like this, but it is only basic logic. This can be illustrated by using dogs as an example (yes, dogs). For example, to enter the category of German shepherds, a particular creature must obey certain rules, and the rules of the subgenre "German shepherd" can never contradict the rules that must first be respected to be a dog. A German shepherd that does not first respect the rules that define what a dog is, is absurd, it is inconsistent logically, it cannot exist.

Similarly, the rules that must be followed to be part of the dog category cannot contradict those that must be respected to be part of the canids category, the rules of canids cannot contradict those of mammals, and those of mammals cannot contradict those of animals... So, a creature cannot be a German shepherd without also respecting all the laws that define these other categories, which are the dogs, the canids, the mammals, and the animals. Each subgenre must necessarily obey the laws of all types that encompass it, in addition to its own particular laws.

What we can do with dogs, we can do with any other concept. Each type that exists in nature can be divided into subgenre, sub-subgenre, and so on. As you get deeper into the divisions, the portrait becomes more complex, but that does not change the fact that every subgenre must *always* respect the laws of all the more fundamental types that encompass it, regardless of the particular laws of its own. Once again, the laws of the subgenres can only be added to those of the more fundamental types; they can never contradict them.

It is the same for the invisible. Surely our brains would explode if we could know all that is possible in the kinds of energy and matter that are invisible to us, so unusual it would be for us. But no matter what kind of energy and matter the invisible worlds are made of, no matter how far they are from what we know today, these will always be just *subgenres* that must respect the great laws that define these fundamental types that are energy and matter.

It is these great laws that are the key to solving the enigmas of consciousness and the origin of life. These laws apply to all forms of energy and matter, and must, therefore, also apply in the invisible since it is necessarily also made of energy and matter. If there are differences, they can only be at the level of subgenres, forms, details... These differences cannot be fundamental.

This allows us to understand, in broad outline, how the invisible is structured since it means that the structure of the invisible is basically the same as that of the visible. The visible and the invisible are only different levels of reality, made of different subgenres of energy and matter, most of which are still unknown to us. This understanding allows us to have a natural and realistic view of the invisible and to move away from the supernatural beliefs that abound on this subject. The visible and the invisible are only degrees, variants of the same thing; there is nothing strange or supernatural about this.

One of the main differences between universalism and materialism is that universalism takes into account *the whole* of reality, whereas materialism takes into account only the visible part of reality. At the heart of materialistic philosophies, there is the belief that most of reality is accessible to our senses and instruments, and that the visible is enough to answer the big questions. For its part, the universalist approach considers that this belief is a huge trap, and it is to avoid this trap that universalism is built on the great laws since the fundamental laws are immutable, while the limits of science are constantly changing.

Universalism considers that most of reality will always escape our instruments, meaning that most of matter will always be invisible and most of energy too. This does not mean that we are condemned to remain completely ignorant about these invisible domains since the principle of universality allows us to have an approximate knowledge of it, to know the main lines, because these main lines are the *same* as those that we find in the visible.

The principle of universality of the laws of nature also enables us to understand that the explanations of consciousness and of life are necessarily in energy, and not in matter; this, just as it allows us to understand that the explanations of consciousness and of life are necessarily in the invisible, and not in the visible.

To arrive at these conclusions, we need to follow reasonings that are similar to those used in the preceding chapter:

1: If the explanation of consciousness is not in matter, it implies that the explanation of consciousness is in energy.

2: The explanation of consciousness is not in matter because the brain is unconscious, as are all material objects.

Conclusion 1: Therefore, the explanation of consciousness is in energy.

1: If the explanation of the origin of life is not in matter, it implies that the explanation of life is in energy.

2: The explanation of the origin of life is not in matter because the first forms of material life are reproductions of a previous life, as are all life forms.

Conclusion 2: Therefore, the explanation of the origin of life is in energy.

Final Conclusion: Therefore, the universality of the laws of nature necessarily implies that the explanations of consciousness and the origin of life are in energy, since any other conclusion requires exceptions to the laws.

The answers are in the universality of the laws, the invisible and energy; and not in exceptions to the laws, the visible and matter. This is how one could summarize, in a single sentence, the essence of the message of this book. For now, we have only seen the basics, but in the chapters that follow, everyone will be able to see the natural solutions available to us when we use this approach.

7.4 IN SUMMARY

An important key to understanding consciousness and life is energy, light in all its known and unknown forms. For those who study the history of science, this is not surprising. Indeed, advances in science have come largely from an ever-deeper understanding of energy and its relationship to matter. Energy plays a central role in the greatest theories of physics, from the beginnings of thermodynamics, at the origin of the Industrial

Revolution, to quantum physics and general relativity, theories in which energy and light have key roles. Since this approach has been so fruitful in the past, why not apply it also to the enigmas of consciousness and the origin of life, by placing energy at the center of our theories?

Once again, it is only a matter of placing concepts in the right place: To solve these questions, we must give energy the *first* role and matter the *second* role. This is because it is the *natural* order of things, an order that science clearly emphasizes through its discoveries.

The primary reason for putting energy at the center of our theories is because it is necessary to respect the universality of the laws of nature. Energy is always what is really important, and matter is always a secondary element. Energy leads the dance, and matter only follows. Materialists have inverted this natural order, giving matter the primary role in the realms of consciousness and life, and then they are surprised when they encounter so much mystery when they try to deepen these questions...

The "mysteries" of consciousness and the origin of life are created artificially by this inversion of the natural order that materialists maintain in their thoughts, just as one encounters false mysteries when one begins to believe the illusion that the Sun is turning around the Earth. We used this analogy when we discussed the importance of the invisible, and we can use it again here because, from a symbolic point of view, the essence of geocentrism is very much like the one of materialism. At the heart of these two systems of belief, we find an inversion of the natural order, introduced when one gives too much importance to appearances.

The Sun is placed at the periphery in geocentrism while the Earth is placed in the center. In the same way, matter is the central element in materialist theories, while energy and light play a secondary role. In the universalist approach, we reverse this logical relationship: We place energy and light at the *center* of our conception of consciousness and life while matter is placed at the periphery, just as the invisible is placed in the center and the visible on the periphery. For the materialists, it is a vertiginous inversion since it is the same kind of mental operation as moving from geocentrism to heliocentrism. But this inversion is necessary to respect the universality of the laws, to treat all things in the same way. Energy and the invisible are the most important parts of nature, they must also be what is more important in our theories dealing with consciousness and life. Only in this way can we build a vision of the world that is truly *coherent*.

In conclusion, here are the primary logical relationships and conceptual unifications we have seen in this chapter:

Energy is the essence of reality.

Inertia is the essence of matter.

Energy is more important than matter.

The universality of the laws encompasses energy and matter.

Energy and matter are levels of reality.

The explanation of consciousness is in energy.

The explanation of the origin of life is in energy.

Again, there is nothing fundamentally new in these solutions since the belief that life and consciousness come from energy has always been with humanity, in many different forms, just as the belief that their origin is in the invisible. Through many belief systems, light is used to symbolize the highest levels of existence, as well as the true nature of the human being. This is an idea that many have intuitively understood, even if, intellectually, there is a lot of confusion on this subject. A confusion that comes mainly from two sources: the supernatural beliefs advanced by religions, and the false interpretations of science advanced by materialists.

To dispel this artificial confusion, one must correctly intellectualize what is already known intuitively, using the most important of all principles: the universality of the laws of nature.

8. CONSCIOUSNESS DOES NOT COME FROM THE BRAIN

Consciousness is light perceiving itself.

The laws of nature are universal, most of reality is invisible, and the essence of reality is energy. We have seen in the preceding chapters why these three affirmations are in perfect agreement with science. We will now rely on these three pillars to build a vision of the world in which consciousness and the origin of life are no longer great mysteries, but phenomena explicable in a natural way.

First of all, let us go back to this idea that has been confusing humanity for millennia—the belief that the answers to these great questions are inaccessible. Among the mistaken beliefs to which we have given power over our lives, it is certainly one of the most harmful. Whoever approaches these questions with this preconceived idea automatically makes the task much more difficult than it actually is.

We all harbor false beliefs that have a negative influence on us. Finding and uprooting these beliefs is an endless job, like protecting a garden from weeds. This work is all the more difficult because many of these beliefs are deeply rooted in our subconscious and are, therefore, difficult to see; while others are inculcated by society, and rejecting them makes one a marginal person, and this may, in some cases, lead to social exclusion and even persecution.

We could extend endlessly on the power of beliefs, a subject that fascinates psychologists, but for the moment, we will be content with a small example to illustrate this power.

Imagine walking alone at night in a neighborhood that is reputed to be dangerous, and suddenly you see someone running toward you. If you believe you are about to be attacked, a series of reactions will automatically follow: your heart will beat faster, you will become tense, you will have cold sweats. Also, you will suddenly be overwhelmed by thoughts that you did not have a second earlier: How can I get out of this situation? Should I run away? Should I call for help? Should I prepare to fight?

All this happens within seconds...until you realize that the person was running just because he did not want to miss the bus that just stopped at the corner of the street behind you! This person then disappears from your life, without ever having been aware of the intense experience that you have just lived, only because you misinterpreted his actions.

It happens to us all, experiencing imaginary fears of this kind, as well as imaginary joys, and these are good examples of the power of beliefs. They show us that as soon as one adheres to a belief with sufficient strength, it gains great power over our inner life. Beliefs can cause all kinds of bodily reactions, as in the previous example, and they influence our thoughts and emotions. No matter whether the object of our belief is real or not, at the level of our inner life, *it becomes our reality!*

It is the same with the belief that the mysteries of life are impenetrable. This belief is widespread; it is part of our cultural fabric and manifests itself in many different forms. Like all beliefs, it influences our mental processes, parasitizing our thoughts on these topics.

We find this belief as much in religions as in materialistic philosophies. On the religious side, it is expressed whenever the supernatural, or "God's mysterious ways," is used to answer a question. Thus, we maintain the belief that the answers to the great questions are in a domain outside natural laws, inaccessible to reason, a domain where everything is possible and that we can fill with the weirdest ideas.

On the materialist side, this belief is expressed whenever we say that consciousness and the origin of life are the greatest mysteries of science, that science is still far from being able to answer these questions, and that perhaps it will never find a solution to those enigmas.

There is a word to describe this practice: it is called *mystification*. To be a mystifier is to make reality appear to be other than it really is. We usually associate mystification with the domain of illusionism, but in reality, there are mystifiers everywhere: in politics, economy, the media, religions, universities...

One of the most widespread mystifications is the idea that the foundations of reality are incredibly mysterious, and that there is only a small elite that can succeed in understanding something about it. With regard to the mysteries of life, this confusion maintains the belief that the answers are accessible only to special people who receive them in the form of revelations or to a handful of scientists working in fields at the cutting edge of science.

In this chapter and the following, we will see a set of solutions to some of the "great mysteries of life." The answers given will all be very simple, to the point of being accessible to children. Because of this, many will be tempted to reject them, following the belief that the answers to these questions are inaccessible or that they must necessarily be out of the ordinary.

Consciousness and life are not mysteries; the labyrinth that surrounds these questions is purely imaginary. This labyrinth exists only in our heads, and not in reality, which operates according to simple laws that everyone already knows intuitively since we experience them at every moment. Our life is built on simple laws, laws that are the foundation of *reality*. So, whoever wants to anchor his or her reflections in *reality* must always remain in the light of natural simplicity to avoid getting lost in the mist of artificial confusion.

We all use simple laws to guide our lives: what is dense tends downward, what is light tends upward; some elements attract while others repel; it is necessary to exert a force to put an object in motion, and an opposing force to stop it; you reap what you sow; two and two always equal four... The greatest mistake is to believe that as soon as we approach the big existential questions, natural simplicity must give way to an obscure logic, which is hardly understandable. This rupture between the natural logic of everyday life and the mysterious logic of the enigmas of life is a false division, that the religious and materialistic philosophies each maintain their own way.

Once again, this division is artificial because it goes against the principle that is at the root of science and at the root of all logical understanding of the world: *the universality of the laws of nature*. The laws manifest themselves perfectly uniformly throughout the universe, completely ignoring the exceptions that our hyperactive brains like to invent. In the past as in the future, here as much as elsewhere, in the visible as much as in the invisible—the same natural logic must apply!

It is the universality of the laws that is the basis of the universalist approach, which is, in reality, a *naturalist* approach since the primary reason for considering that consciousness and life do not come from matter is that this is the only way to explain them using *natural* solutions, that is, using the same simple laws that we use in everyday life. To be able to use the same solutions, as much for everyday problems as for the great existential questions, is not "simplistic," but rather another consequence of the universality of the laws.

Simplicity is a *force*, not a weakness! The mystifiers like to bury the subjects of consciousness and the origin of life under a heap of complex considerations to justify why they are unable to give us clear answers. But the approach we will use follows the opposite way: it only accepts what fits with natural logic, what is as clear as "2+2=4". It is the only path that leads out of the mist of artificial confusion.

8.1 MOST OF THE HUMAN BEING IS INVISIBLE

Now, let us look at the enigma of consciousness and see how it is possible to solve it using only solutions in accordance with the laws of nature.

As a starting point, we must, first of all, consider this concept: the idea that the human being, in addition to the body, also has an invisible part, inaccessible to our senses and the instruments of current science.

This concept is very simple and natural, and yet, there are many, especially in the scientific community, who see it as an irrational proposition. For these people, the idea that the human being has an invisible part is too similar to the beliefs conveyed by religions for them to give it credibility. The resistance that this idea encounters is one of the most glaring examples of the power of materialistic beliefs in scientific circles—all this because most researchers believe that considering consciousness as a mysterious product of the brain is the only rational option.

This is false, of course, because since much of *nature* is invisible, to consider that a part of the human being is also invisible is a *natural* solution, and therefore, an acceptable hypothesis from a scientific point of view.

Considering that a part of the human being is invisible is not unscientific because this idea does not contradict science, only certain materialistic dogmas. On the contrary, this idea fits perfectly with the discoveries of science, especially those of physics, which show that the part of reality that we perceive is very small compared to all that exists. So, again, the materialists install artificial divisions, since they accept the idea that most of nature may be invisible, but consider as irrational the idea that the human being can also have an invisible part.

If we do not find the answer in what we see, it is simply because the answer is in what we do not see! The first success of this approach is that it allows us to completely eliminate one of the greatest mysteries of current science, the enigma of the process that is supposed to allow the brain to generate consciousness.

Indeed, if we accept the idea that the process of consciousness does not take place in the brain, it means that the brain is not conscious. If the brain is just as unconscious as any other material object, this enigma disappears since it is only an artificial problem generated by the belief that the brain is conscious.

The first step in getting out of the confusion is, therefore, to consider that what we perceive of the human being, its body, is not the whole of what constitutes a human, just as what we perceive of nature is not the whole of what constitutes nature, and to consider that it is in this invisible part where consciousness resides. Contrary to popular belief, seeing things this way does not bring these reflections into the realm of the paranormal, but rather, *it allows us to insert the human being into the natural order*, where the invisible is much more important than the visible.

It is by neglecting the invisible that we disconnect ourselves from reality, and so, to reconnect with reality, we must put the invisible at the center of our vision of the world. This, never forgetting that the invisible worlds are like the visible worlds, that is to say, domains governed by the laws of nature. The main difference between the visible and the invisible is that the invisible worlds are made of substances poorly understood by present-day science.

In other words, it is quite rational to believe in the existence of invisible worlds if we consider these worlds as part of nature. What is irrational is to believe that these worlds are supernatural domains that escape logic, as religions do; or to believe that these domains do not exist or that they are of little importance, as the materialists do.

Here, represented as a diagram, is the distinction between the visible and the invisible parts of the human being (next page). Even if, for simplicity, the invisible part is represented as a single element, it should not be conceived as uniform, but as a living structure made of a great variety of substances and sub-elements, because the invisible worlds are as rich and complex as the visible worlds are.

To solve the mystery of consciousness, the first step is to accept that the center of our consciousness is situated in the invisible part of the human being. Instead of the "center of consciousness," we could also use the terms "place where consciousness resides," "organ of perception," "conscious substance"... This can also be called "the spirit," if one is willing to move away from the supernatural beliefs that usually accompany this word. In the context of this book, the word "spirit" simply designates the part of the human being that is conscious, its center of consciousness, which is also its center of will.

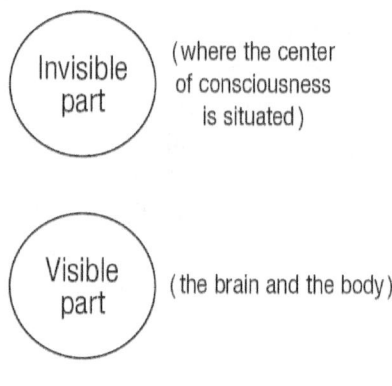

Regardless of the labels, what matters is that this point is not located in the body. This allows us to emphasize the most important logical relationship to be grasped: the understanding that consciousness is *independent* of the brain, and that it resides in a center that has its own existence on the invisible side of reality.

For materialists, this represents a step backward in our understanding of the human being because it is similar to the notion of soul or spirit that has accompanied humanity since the dawn of time. Materialism is born with the abandonment of these notions, with the belief the human being is only the body. They see this as "progress," even if this approach has led them into logical dead ends from which they try to escape by adhering to their own superstitions, such as the belief that the brain has mysterious powers that no other object possesses.

The first step in solving the enigma of consciousness is, therefore, to consider that the center of consciousness of the human being, the spirit, is independent of the body and is situated on the invisible side of reality, a part of nature from which we are still far from grasping the real importance.

From this point of view, the brain is nothing more than what experiments tell us it is: an object that is not fundamentally different from any machine made by humans. The brain is an object like the others, it has no special power; consequently, to search in this organ for the origin of consciousness has no more meaning than to go hunting for unicorns. For generations, scientists have analyzed the brain in all kinds of ways, without ever discovering in the structure and functioning of this organ anything particular to explain the imaginary powers that materialists think it has. It is clear that brain cells, neurons, are specialized in the processing and transmission of information, but how does this explain the emergence of consciousness? For the materialists, the mystery remains whole. All that scientists have managed to do so far, using different brain imaging techniques, is to *correlate* brain activity with certain states of consciousness without ever being able to explain how the brain could *cause* consciousness. How much longer, before they realize that it means that the brain is only *in relation* with consciousness and that the cause of consciousness is elsewhere?

One of the first lessons of science is that one must always be careful not to confuse what is only a correlation with what is a cause and effect relationship. The fact that some patterns of neural activity accompany certain mental processes, sensations or states of consciousness, is not proof that these are caused by neurons—all this proves is that there is a relationship between the brain and consciousness.

Those who jump to conclusions, considering such observations as proof that consciousness comes from the brain, show a lack of intellectual rigor. Indeed, in science, it is not permissible to draw a definitive conclusion before having rigorously considered all the possibilities, that is, before having explored all the natural solutions. As we have said before, placing consciousness on the invisible side of nature is a natural solution since parts of nature are invisible. Therefore, when materialists refuse to seriously consider invisible solutions, by sweeping them under the big carpets of the "supernatural" or the "paranormal," they show a dogmatic behavior that has no place in science, where invisible solutions are allowed.

A big part of the history of science is the progressive discovery of all that is contained in the invisible! Indeed, every time a new instrument was invented that allowed us to explore areas that were previously invisible; such as the microscope, the telescope and the radio; researchers were amazed by all that they discovered in these areas, and they also found answers to many questions that previously seemed unsolvable. In spite of all these advances, so far, science has only touched the surface of reality. An inconceivable richness is still waiting to be discovered on the invisible side of nature, a domain that encompasses most of existence and that contains the answers to so many questions that humanity asks about consciousness and life.

8.2 THE BRAIN IS AN INTERMEDIARY

By considering the brain only as an *intermediary* between the invisible domain and the visible domain, the universalist approach eliminates some of the mystery of the brain. On the other hand, it brings a new question: How are the center of consciousness, located on the invisible side, and the brain, located on the visible side, able to communicate with each other?

For the spirit to express itself through the body, as well as to perceive the world through it, there must necessarily be an interaction between the two, an exchange of information. At first sight, this exchange may seem very mysterious, and one would be tempted to see there a subject to which science is still far from being able to provide an answer. But, again, this answer is much simpler than what is generally believed.

The key to solving this enigma is the same key that can solve *all* the great enigmas: the universality of the laws of nature. According to this principle, the laws act everywhere in the same way, both in the visible and the invisible, and so, if one wonders how the spirit and the body can interact, to find answers one only needs to observe how interactions occur in general.

Interactions are governed by a great law, which can be called the law of interactions: *Every interaction is an exchange of energy that carries information.* This very broad definition of interaction encompasses an incredible variety of phenomena. For example, there is speech, in which information is transmitted by sound, a series of delicate airwaves that transmit impulses to hearing organs located in the ear. There are also the images that appear on our computer screens, which are sequences of light pulses transmitted by the pixels that make up the screen. There is also the Internet, where extremely fast and precise impulses transmit information traveling in wires or via electromagnetic waves. Another example

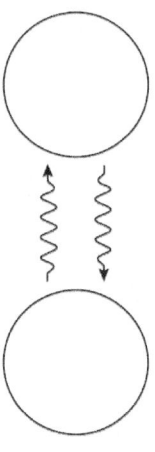

This universal pattern of interaction represents any interaction between two elements, including the interaction between the visible and the invisible aspects of the human being. The arrows represent an exchange of energy, and the waves represent the sequences of pulses into which the information is encoded.

is, of course, neurons, where a series of electrical and chemical impulses transmit information to the sensors ready to receive them, triggering all kinds of processes.

The interactions can take an infinite variety of forms, but, whatever the form, the principle remains the same: All interactions are exchanges of energy that carry information.

For the one who accepts this law, the question of the interaction between the spirit and the body appears immediately less mysterious: no matter the form of this interaction, it can only be a transmission of energy carrying information. So, we can also use the previous diagram to represent the interaction between the visible and invisible parts of the human being or between the center of consciousness and the body.

We have taken a step further in our understanding, but this diagram brings another question: What is the medium of this exchange? To answer this question, we must, once again, turn to the greatest ally of reason: the universality of the laws of nature.

To solve the question of the interaction between spirit and body, we explored how interactions occur in general. To solve the question of the medium, we must do the same thing, and question ourselves about what is, in general, the medium by which interactions occur.

The answer is already contained in the definition of the word interaction: exchange of *energy* that carries information. The medium of the interactions is *always* energy, and it must be the same thing for the spirit and the body; their interaction must be conducted through energy.

Energy is a very large concept. To make this answer more concrete, we can replace the word "energy" with the word "light." Because the interactions, in all their forms, can be seen as exchanges that are made through light.

It should be remembered that the word "light" is used here in a broad sense, it designates a whole family of particles. In the context of this book, this word does not only mean the light that our eyes perceive, the electromagnetic waves, the particle of which is the photon; this word also refers here to the elementary particles of the same family as the photon, which physicists call "elementary bosons," the vehicles of all the known interactions of physics. In addition to this, the word "light" refers here to all the particles of this family that are known, but also those that are still to be discovered. In short, inside this book, this word refers to one of the two large families of elementary particles that constitutes nature: the bosons, the particles of energy or force. The other family being the fermions, the particles of matter.

One might think that such a broad definition is not useful, but on the contrary, this definition is extremely effective in avoiding artificial confusion. When one always keeps in mind that there are only *two* families of particles, everything becomes clearer since instead of conceiving the universe as a complex mixture of particles with strange names, we can conceive of it with only two kinds of substances: energy and matter. There are a large number of known and unknown particles in both families, but the basic behavior remains the same: the identical particles of light can interpenetrate, whereas, for identical particles of matter, this is impossible.

We have already seen these notions in Chapter 7. In science, words are only labels. If the word "light" is used here, in a sense that is not necessarily that of conventional science, it is because this word allows us to understand *intuitively* what this kind of substance is, much better than specialized terms such as "bosons" allow us to do. Because we all have the *experience* of light, we have all seen rays of Sun piercing the clouds, and even the blind feel the warmth of the spring, the one that makes the flowers bloom!

Light is the energy that animates everything. The universe is an ocean of light, it is present everywhere. Between every atom, every planet, every star, space is filled with particles of light that convey *all* the interactions. This is what physics tells us: fermions can only interact through bosons, which are particles of the same nature as light. Therefore, it is impossible for an element to interact with another element without this interaction passing through light in all its forms. It is one of the greatest discoveries of physics, and it is a notion that gives us a clear vision of how the universe works. It is the movement of light that leads the dance in the universe; matter is inert and only obeys the impulses it receives.

All exchange of energy passes through light since energy and light are, in fact, the same thing. This is also the key to understanding the relationship between spirit and body: they can communicate with each other only through light since it is the *universal mediator*.

All around us, we can see the wonders that can be achieved by exchanging information with light. Cellphones, satellite telecommunication, wireless Internet...our entire civilization depends on these exchanges that are made through electromagnetic waves, through a form of light at frequencies invisible to our eyes. All these examples show us how it is possible to control a wide variety of devices using light, and it is through this process that we can explain how the spirit controls the body, which is basically an *instrument* like any other.

This process is fundamentally simple, and it can be illustrated with a familiar example: a remote-control device. Consider a vehicle remotely controlled by a pilot. It can be a small remote-controlled car driven by a child or a military drone piloted from a control station located thousands of kilometers away; regardless of the form of the device, the principle remains the same. This is a system that we can illustrate using the universal interaction diagram previously seen.

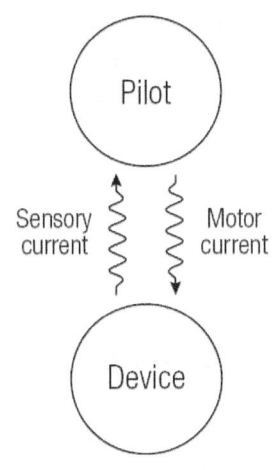

Here, the information streams have been labeled. The first current is that which goes from the pilot to the vehicle, transmitting orders to it. It can be the waves emitted by the remote control in the example of the remote-controlled car or transmitted by satellite in the example of the drone. Since this flow of information is intended to indicate to the device how it should move, what actions it must perform, we can call it the "motor current."

The second current is the one flowing in the opposite direction, from the vehicle to the pilot. This current provides the pilot with information about the environment of the vehicle as well as the status of the vehicle itself. It is addressed to the senses of the pilot, mostly sight and hearing, and can, therefore, be called the "sensory current." In the case of the remote-controlled car, this current of information is perceived by the simple observation of the vehicle, whereas, in the example of the drone, this current is transmitted by satellite, and the pilot perceives it by means of screens. This second stream informs the pilot of the consequences of the orders he has sent, and based on this information, the pilot can choose to modify the route taken by the vehicle by sending it new instructions.

Therefore, two information streams are flowing in opposite directions from each other: the motor current and the sensory current. This operating principle, where two opposing currents do complementary work, is another universal pattern. It is found everywhere: first of all, in the human body, where the neurons that transmit the motor information, that is to say, the ones that activate the muscles, work in parallel with the neurons that transmit the sensory information perceived by our senses. This is also the case for the blood circulation, where the arterial current, which brings the blood from the heart to the tissues, is separated from the venous current that brings the blood from the tissues to the heart. Similarly, in our homes, drinking water does not circulate in the same circuit as wastewater. We could list many other examples: sap circulation inside plants, atmospheric and oceanic currents, electrical circuits, car circulation…

In all these cases, we find this separation in *two complementary currents*. It is a principle that is extremely simple and natural, and that can explain many things since it is found at the base of the functioning of nature. This circulation of two complementary currents is a consequence of another great law of nature, the law of retroaction: *Every action leads to an equal and opposite reaction.*

Therefore, it is through a constant flow of action and retroaction, conveyed by two complementary currents, that the remote-controlled vehicle is guided by means of electromagnetic waves, that is to say, light. It is this very simple operating principle that can enable us to understand the relationship between the spirit and the body, which can be seen as any other relationship between a worker and his tool, an artist and his instrument, or a driver and his vehicle.

Materialists believe that the human body is more than only a vehicle, and this belief gives rise to many artificial mysteries. This belief is comparable to a situation in which a person would meet for the first time a remote-controlled vehicle, without knowing that such things can exist. Faced with the behavior of this machine, which seems to react sensitively and intelligently to its environment, this person could jump to conclusions and consider that it is a conscious machine. On the other hand, if this believer decides to go further and dissect the machine in search of the source of this conscious activity, he will not find it inside the machine. He will only find pieces intended to capture, process and transmit different currents, but no "consciousness-generating machine."

It is the same situation in which materialists find themselves when studying the human body. By analyzing the body, they find only machinery: channels, tanks, valves, pumps, filters, sensors, switches, computers… Systems similar to those that humans are capable of making themselves, only much more refined and complex. When materialists seek the source of consciousness in the body and the brain, they find themselves in front of a "big mystery" because when they study them, they find no difference with the other machines that they themselves regard as unconscious.

For those who look at this in an objective way, the conclusion is clear: we must consider the brain as being as unconscious as the other objects because there is no objective difference between them. This conclusion does not create any artificial enigma since, to explain the manifestations of the will and consciousness, we have another solution, which is quite natural: they manifest themselves in the body only through *transmission.* That is, through energy currents, in the same way that any unconscious vehicle or tool is controlled. The principle is exactly the same, only the form is different. There is nothing esoteric here. The solution is trivial, and it operates according to laws that we see at work everywhere around us.

In the universalist approach, the body is *nothing more* than a vehicle, a tool, an instrument. If we have the impression that the relationship with our body is different from the relationship we have with other tools, it is because the connection between the spirit and the body is very intimate, to the point that it gives us the impression that we are our body.

But this illusion can also be easily explained by the analogy of the remotely controlled vehicle. To explain this, we must go further in this analogy by imagining that instead of an ordinary vehicle, the remote-control device is a robot possessing a human form; and instead of being controlled through remote controls or screens, it is controlled by a sophisticated virtual reality system.

We can imagine that the pilot perceives, using a virtual reality helmet, the sight and sound of the environment in which the robot walks, but also, that the pilot is wearing a suit that covers his entire body, which transmits to his skin all the sensations that are received by thousands of sensors spread on the surface of the robot. It is easy to imagine that by multiplying the sensors, by making this system ever more sophisticated and precise, we can produce, in the pilot's mind, the *illusion* that he himself is the robot. Then, we only have to leave the pilot in this illusion long enough to see him forget that he is, in reality, something other than this machine.

If we have the illusion of being our body, it is because the level of *connection* between it and our spirit is very high. We are constantly inundated with information from millions of sensors located inside the body and on its surface. This way, we receive all kinds of impressions which, once united, form an illusory image of ourselves, which is superficially very convincing—but it remains an easily explicable illusion.

The body is an extremely sophisticated instrument, but an instrument all the same. It cannot do anything more than any other tool can do, that is to say, receive instructions from the motor current, process them, and transmit the results via the sensory current. This interaction between the spirit and the body is invisible, just as the interaction between the pilot and the remote-control device is to our eyes, but it is not magic because it operates according to the same natural laws.

In the case of the body, one can conceive that certain parts of the nervous system can act as antennas, intended to capture the instructions that the spirit transmits, instructions that the brain must translate before transmitting them to the rest of the body. One can conceive that there is also an opposite current, where the information perceived by the senses is transmitted to the center of consciousness after having been collected and processed inside the brain. Here, it does not matter via what types of waves or particles, known or unknown, this communication is conducted; it is also irrelevant whether the spirit is conceived as being inside the body, like a driver in a car, or outside, as in the case of a remotely controlled vehicle. *What matters is understanding the operating principle,* not the exact form. It is a solution that operates according to the same rules as any other system. No supernatural or paranormal explanation is needed; it is only transmitters, translators and receivers, among which information circulates.

The processes taking place inside the brain are extremely complex, and only two categories of activity are presented here in broad outline: the treatment of the motor current and the treatment of the sensory current.

On the one hand, the brain transforms the motor current coming from the spirit, which, in this case, can be called the current of the will, adapting it to material reality. The spirit has objectives, but since matter is foreign to it, it cannot, on its own, understand how to achieve these objectives in the material world. It is as if the spirit and matter speak different languages, so the spirit needs the help of a translator: the brain. Instead of the term "translator," one could also use the more technical term "coder-decoder," which is the same thing. In short, the brain receives instructions from the spirit through the current of the will, decodes them, encodes them into a new language and sends these freshly translated instructions to the different parts of the body intended to apply them.

It is a work of *division* since the brain receives a unique will, a unique goal, and divides it into many actions, many steps. For example, take someone who wants to become a musician. This is a unique goal: "I want to be a good musician." It is wonderful, but wanting it alone is not enough, one also needs the means. It is the role of the intellect, the ability to think coming from the brain, to determine what steps to follow to reach this goal: buy an instrument, take classes for years, practice thousands of hours... To realize this unique will from the spirit, the intellect must translate it into a complex succession of steps adapted to material reality, through a lot of planning.

In other words, the spirit gives the *drive* and the intellect deals with the *means*. The spirit provides enthusiasm, and it is the role of the intellect to channel that energy to ensure that it is not lost but does an effective job in the material world. To do this, the intellect uses its mental representations of reality, which are maps it has developed over time, diagrams that show which are the good and the bad channels into which it can send this energy, according to our goals. In this sense, the intellect can be seen as a navigation system that the spirit uses in its journey into matter.

Hence, the importance is ensuring that our mental maps do not contain too many errors due to misinterpretations or wrong beliefs because bad models can lock the spirit into a mental prison, and its energy will be diverted to erroneous ways. The seriousness of this problem cannot be overstated since all misfortunes have their roots in the wrong representations and false beliefs that we cultivate, which have become a dense jungle that stifles the spirit and prevents it from flourishing.

This is, in broad strokes, how the brain works with the current of the will coming from the spirit. Let us now see how the brain works with the sensory current, which can also be called the current of consciousness.

The current of consciousness flows in the opposite direction of the current of the will, so the work done by the brain is the opposite of the previous work. Previously, it was about dividing the will of the spirit into a long chain of actions; for the current of consciousness, it is more about *uniting* the information coming from our different senses to build a coherent image of our environment, an image that is then transmitted to the spirit.

As mentioned before, millions of sensors are scattered everywhere in the body and on its surface. Each of these microscopic sensors is sensitive to a different type of information: those of the retina are sensitive to light, those of the ear to the vibrations of the air, those of the tongue to chemical substances, those of the skin to pressure…

Despite the wide variety of these sensors, they all work on the same principle; when a stimulus exceeds a certain threshold, the sensor triggers and sends signals to the brain via the nerves. In this way, the brain is constantly inundated with millions of signals, an incredible amount of information that it has to deal with. This amount of information is so huge that it forces the brain to prioritize. The signals that are deemed useful for our current activity are put in the foreground, while the others stay in the background, and we are hardly aware of them. For example, even if our little toe is constantly sending signals to our brain, this information rarely comes to our consciousness. Except when we hit this toe against a piece of furniture; then, all the other signals go into the background, and our little toe becomes all that there is in our mind!

Therefore, the work of the brain on the sensory stream is to process the enormous amount of information coming from our environment, from which the brain extracts an image elaborated according to our priorities of the moment. It is only this thin slice of reality that we perceive consciously; the other processes remain unconscious.

As for the intellect, while for the current of the will it has to do the work of *planning*, for the current of consciousness, it has to do the work of *interpreting*. Through this work, the intellect uses the information we receive from the environment to construct theories, representations of reality, establishing the logical relationships between the elements we perceive. In this way, it develops mental maps, these same maps that it then uses to plan, predict and explain. Once again, this is delicate work since our senses convey a very fragmentary vision of reality, so we must constantly be careful not to jump to conclusions until we have accumulated sufficient information; otherwise, our theories may be wrong. In particular, our senses are vulnerable to illusions of all kinds, such as the illusion that the Earth is flat and that the Sun revolves around it. On the basis of illusions of this kind, the intellect has, in the past, constructed false representations of reality to

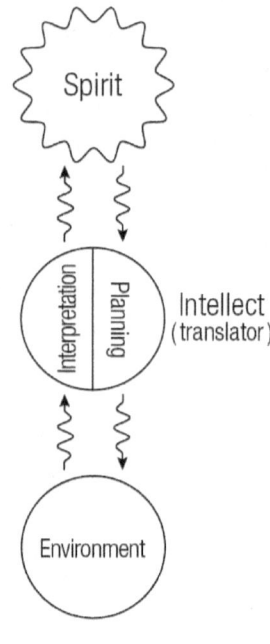

The intellect is an intermediary between the spirit and its material environment. It does planning work on the current of the will and interpreting work on the current of consciousness.

which many people have adhered for millennia. To progress, it is necessary to add another illusion to this list of errors: the illusion that will and consciousness have their origin in the brain.

The work of the brain on the currents of the will and of consciousness has been summed up very succinctly here. In reality, these are incredibly complex processes that a brain specialist can study for a lifetime without grasping all their details, just as a biologist can study a single cell for a lifetime and constantly discover new things—nature's richness is inexhaustible. Nevertheless, these broad guidelines allow us to have an intuitive understanding of these processes.

This model, in which the brain is considered only an information-processing center, agrees with scientific knowledge about the functioning of the brain. Indeed, all the studies that have been conducted on this organ show us that it specializes in the reception, analysis, storage, translation and transmission of information. For this reason, it is often compared to a computer, which is the man-made machine that is the closest to what a brain is.

The observations that scientists have made about the brain are far from being superficial. For generations, they have analyzed it in all kinds of ways, up to the level of molecules. They have accumulated a phenomenal amount of data about it, without ever discovering *anything* that could lead us to believe that it is something other than an information-processing center. As the stomach absorbs food, as the lungs transform air, as the heart pumps blood, the brain manages information! Despite this, materialists continue to believe that it is more than an organ that only processes information, that it can do something more, namely "generate consciousness."

This confusion comes mainly from the fact that it is possible to modify our conscious experiences by acting on the brain, just as it is possible to modify how we can express our will, modifications that materialists interpret as proof that the brain is the source of consciousness and will. But, again, they are just jumping to conclusions.

By acting on the brain, it is indeed possible to modify what we feel, as well as how we can express ourselves, whether through surgery, accidents, illnesses, drugs, electrode stimulation or even changing our beliefs.

All this can have effects, the wide variety of which astonishes researchers: language disorders, dyslexia, epilepsy, Parkinson's disease, Alzheimer's disease, memory disorders, autism, hallucinations, schizophrenia, depression, coma, paralysis... The list of the consequences that the disorders of the brain and the nervous system can produce go on indefinitely.

But these observations are entirely in agreement with the idea that the brain is just an intermediary; only, it must be understood that it is an *active* intermediary. It does not transmit passively, like a plumbing pipe. On the contrary, it does very complex work on the information that passes through it, as has been explained previously. Therefore, when it is altered, the brain is no longer able to perform this work in a normal way, which modifies the image of reality that it transmits to the spirit via the current of consciousness, or modifies how the spirit can express itself through the current of the will.

To understand these phenomena better, let us make an analogy with the functioning of the eye. Based on superficial observations, one could easily conclude that the perception of images must occur in the eyes; in the same way that materialists conclude that perceptions, in general, occur within the brain. Indeed, the illusion that the perception of images is made at this point in our body is convincing. If we close our eyes, we no longer perceive images; if the eyes are altered because of diseases such as myopia or presbyopia, our perception of the images is blurred; if it is the retina that is affected, it is our perception of colors that is modified, as with color blindness. All these observations could lead us to conclude that the perception of the images must be made in the eyes since when the eyes are altered, the images are altered too.

Of course, we all know that this is not the case. These observations are only proof that the eyes are *related* to the perception of the images, and not that they themselves perceive or feel the images. In reality, the eyes only receive and process visual information before transmitting it to the brain via the optic nerves. Inside the brain, specifically in the occipital lobe, located at the back of the head, the processing of visual information continues. Just as it is possible to change the perception of the images by acting on the eyes, it is possible to change it by acting on this region of the brain. People with problems in this area may experience visual hallucinations, vision problems, and even blindness, all this while their eyes are in good health.

For the materialists, this is enough to convince them that the perception of images must be in the brain itself while the non-materialists remain skeptical. Indeed, what proves to us that this is not another illusion, just as the impression that the perception of images happens in the eyes?

The only answer that the materialists can give us is to say that the perception of the images must happen in the brain since, according to them, there is nothing next. They follow the same reasoning for all perceptions:

it is in the brain that the nerves end, it is the end of the path, and therefore, the conscious perception must happen there. But when the materialists try to understand what is happening in this organ to explain the phenomenon of consciousness, this "end of the path" quickly becomes a logical impasse, since they cannot do it. All they have to offer us are fuzzy explanations, magical thinking, and "great mysteries." All this is mixed up in the cult of the brain that materialists propagate, after taking good care to wrap everything in scientific vocabulary, so that it looks more credible, just like science fiction gives us fantasy stories presented under a cloak of science.

For their part, non-materialists do not let themselves be impressed so easily. No matter how convincing the appearance is that the perceptions take place in the brain, from a scientific point of view, *it has no weight.* Is it not one of the primary lessons of science that we need to be wary of appearances?

We know that an important part of reality is invisible, so nothing forces us to believe that the circulation of information stops in the brain! On the contrary, everything becomes clearer as soon as we accept the idea that the path continues beyond the brain, and that from this point, information travels in invisible ways. In other words, to get out of the dead end, we must accept the idea that the circuits of the brain and the nervous system are connected to other circuits, which are invisible, and which lead to the invisible part of the human being.

As already mentioned, this solution eliminates the so-called mystery of the brain, since, in this process, it is nothing more than an information-processing machine, exactly what experiments tell us it is. In addition, seeing the brain as an active intermediary, a translator, allows us to explain how it is possible to significantly change our perception and our ability to express ourselves by acting on it.

Of course, adding invisible circuits and elements, working in concert with those that are visible, adds a layer of complexity. But, that the structure of the human being is actually much more complex than our superficial impressions let us believe is entirely consistent with science.

Nature is incredibly complex, scientists cannot even describe all the processes taking place inside a single blade of grass! Whether in cells, organs or ecosystems, the complexity of interrelations, the refinement of all the structures, far exceed our understanding. Is it any wonder then, that the human being is actually much more complex than the materialists believe? No, on the contrary, it fits perfectly in the natural order.

It is not surprising either that the solutions are invisible, scientists should be the first to find this quite normal, they who keep saying that one of the greatest contributions of science to humanity is the understanding that our senses or instruments perceive only a tiny part of reality.

All this complexity and these invisible elements could discourage us from addressing these issues, believing that it is impossible to see clearly. But, fortunately, there is a master key that we can use to clarify everything: the universality of the laws of nature. Because, no matter the level of complexity of the elements, no matter whether they are invisible or not, the laws always act the same way!

The structures can be very complex, *but the laws are always simple.* It is on these laws that we must focus if we want to build a clear representation. That is why we use simplified illustrations in this book, which focus only on operating principles, not details. As soon as we look at the details, we discover an infinite variety of phenomena, which, even if they are fascinating, can make us forget that all this is driven by the same simple laws.

It is the same for the so-called mystery of consciousness, the solutions provided by the laws of nature are, in fact, very simple. The body is only an instrument, operating according to the same principles as any other instrument. It is a vehicle that is controlled, using information-carrying energy currents, by a pilot who is invisible: the spirit, the center of consciousness and will.

8.3 THE LAW OF INERTIA

There is a very simple reason explaining why it is impossible for consciousness to come from the brain, and this reason is another law of nature: the law of inertia.

The law of inertia is one of the primary pillars of physics. Like all natural laws, it can be formulated in many ways, but it can be summed up simply as: *Every form of matter only resists change.*

Inertia can be defined as *resistance to change*. This means that an object tends to remain in the state that it is: if it is at rest, it will remain so as long as a force does not compel it to move, and if it is in motion, it will continue its movement in a straight line as long as a force does not compel it to change direction or to stop.

This is very common knowledge. We all know that a stone is not going to suddenly start moving unless it receives an impulse and that once thrown, this stone will not suddenly stop its course unless it meets something on its way. We also know that the more material an object contains, the more massive it is, the more it will resist before being set in motion, and the harder it will be to stop once in motion.

Inertia is an intrinsic property of matter. The two are inseparable—the law of inertia is the law of matter. From a certain point of view, we can even say that matter *is* inertia, just as light is energy. Therefore, particles of matter can be seen as "particles of inertia," or "points of resistance," and all the behaviors of matter essentially come from its ability to resist change. This change comes from the circulation of energy particles, which, unlike matter, have their own motion and are, therefore, constantly moving without the need for external impulses. The particles of matter resist the circulation of energy, creating all sorts of constraints, and the particles of matter also exert resistance to each other, obeying the famous Pauli exclusion principle.

All natural phenomena result from these exchanges between energy and matter, with energy on the one hand, which drives all things in its movement, seeking to distribute itself as uniformly as possible, and matter, on the other hand, resisting this movement. All the structures of nature are born of a balance between these two primary behaviors of the universe, which are complementary, like fire and water.

This is very interesting, but what is the connection between inertia and the subject of consciousness? Well, *there is none,* and this is a big problem for the materialists. In short, the problem at the heart of the materialist approach of consciousness is that it considers that consciousness can be born of matter, while the laws of physics tell us that matter *never* does anything other than be inert!

Materialists believe that by combining many inert elements into complex structures such as the brain, one can reach a magical threshold where these elements cease to be just inert things, to start doing something more, that is, to "generate consciousness." But when asked what laws of physics explain this extraordinary phenomenon in which they believe, they are very confused, since the only thing matter can do is be inert.

The materialist's favorite explanation is usually to say that consciousness is an "emergent property" derived from the complexity of the brain. But how does the complexity of the brain differ from that of other complex objects that are unconscious? This is a mystery for the materialists.

Matter can be organized in *networks of resistances* that can be extremely complex, and these networks can adopt all kinds of amazing behaviors.

However, under no circumstances can these networks begin to generate consciousness, because, in reality, they always do the same thing under different forms; that is to say, capturing energy and redistributing it in various ways.

For example, the millions of transistors that are inside a computer chip are just forms of resistance, and by manipulating the activity of these resistances with programming, we can produce countless different results. This can be interpreted as saying that new behaviors emerge from matter with each new program, but this interpretation only works at a superficial level. At a fundamental level, the behavior of matter remains strictly the same: it only resists change. The only thing that changes is the order of activation of the resistances inside the network. It is the same thing for the matter that is inside living organisms. It is no different from matter found in non-living objects; it has only been *programed* by the genes to behave in a special way. Living matter is only programed matter.

Once again, it is the universalist approach that best matches the discoveries of science. Indeed, physicists have studied matter for centuries, and the behavior of matter is well understood through the many formulas of physics. Throughout these formulas, matter remains a passive element, which only obeys, with more or less resistance, the instructions it receives from impulses. There is nothing mysterious about it, all this is well known. The universalist approach only follows what physics tells us: since matter is fundamentally inert, it must, therefore, always be considered an inert element. It does not matter whether this matter is in a stone, a computer, a brain, or a bowl of spaghetti; it remains an inert element that never does anything but react passively to the impulses it receives from energy.

The answers are in the *BASICS* of science—it is impossible to insist on this point too much. The answers are not in believing in inexplicable fantastic phenomena, but in a correct interpretation of the fundamentals of science, an interpretation free from the materialistic beliefs that confuse everything. The law of inertia alone suffices to refute materialism since it tells us that matter is always inert; therefore, those who believe that it can do something more only attribute imaginary powers to it.

Of course, the materialists will reply by saying that to believe that consciousness comes from energy is not a better solution, since, according to them, it is necessary, then, to give imaginary powers to energy rather than to matter. However, when we think about it more deeply, we realize that the notion of consciousness fits very well with the notions of energy and light, as we will see in the following section.

8.4 CONSCIOUSNESS COMES FROM ENERGY

The existence of consciousness in a center independent of the body makes it possible to solve many questions that haunt current science. On the other hand, skeptics will say that it only displaces the problem. Materialists are unable to explain how consciousness is born in the brain, but are non-materialists able to explain how consciousness is born in the spirit? This interrogation is another opportunity to demonstrate the enormous explanatory power of the universalist approach because this approach also makes it possible to provide answers to this question.

To see more clearly about the question of consciousness, we must first ask ourselves this other question: "What can we perceive?" Indeed, since we can define consciousness as the *ability to perceive,* to understand the nature of this capacity, we must first understand what the objects of our perceptions or sensations are.

This is a key that is generally overlooked by those who question the mechanism of consciousness, because the answer to this question may seem very ordinary. What we perceive is what is in our environment: tables, chairs, walls, the sky, trees, houses, our neighbors, apples, or whatever... The objects of our perceptions seem, at first glance, very disparate; yet, when we think more deeply, we realize that in reality, the object of our perceptions is always the same: *light.*

Indeed, when we look around us, we do not directly perceive the objects themselves; we perceive *images,* that is to say, the imprints that these objects left in light. When we look at an apple, it is not the apple we see, but the light that is transmitted by the apple. In reality, it is not the apple that is red, it is the light! In itself, the apple has no color because, when light is removed, it becomes black, like all material objects.

We never perceive objects directly, we only perceive the light they transmit. This rule applies not only to sight but also to all our other senses. Let us remember that the particles of matter, the fermions, never touch each other directly; they interact only through the particles of energy, the different types of light, the bosons. When we take the apple in our hands, when we hear the noise it makes when we take a bite from it, when we taste its flavor...in all these cases, what we perceive is light. Because it is impossible for the material of our hands to touch directly that of the apple, it is impossible for our organs of hearing to touch the air, which conveys the sound waves, and it is impossible for our tongue to touch the bite that we chew. In all these cases, it is actually light waves that are exchanged between us and the elements we perceive. These waves permeate all things, they are constantly in circulation, conveying an inconceivable wealth of information. Ultimately, it is this information contained in light that we perceive—and nothing else.

We do not perceive matter, we only perceive light. No one has ever seen matter, touched matter or tasted matter. It will always remain something *foreign* to us, that we can perceive only indirectly, through the light.

Scientists seek consciousness in the meanders of the brain, without seeing that there is, in these discoveries of science, an important key to understanding the nature of consciousness. For, if light is the only thing we perceive, it indicates that our center of consciousness, our organ of perception, must necessarily be of the same nature as light!

Only light can know light; it is a reality anchored in the laws of physics, in the fact that only the bosons can interpenetrate. This means that if our spirit is made of a substance of the same nature as light, it has the ability to be "one" with the information it receives through it. The perception would then arise from the very intimate union between the spirit and the information it receives, a union made possible by the fact that they both exist in the light, the energetic part of reality.

Saying that the spirit has the ability to perceive the world because, in some sense, it can become "one" with the information it receives, only gives us an overview of a process that certainly contains many complex steps, like all natural processes. But the understanding that it is a process that unfolds *entirely in energy* already allows us to see more clearly, even if this understanding must primarily remain at the level of intuition.

If the spirit is made of energy, of a particular kind of light, and perceives the world only through light in all its forms, a definition of the consciousness that can be formulated is this: *Consciousness is light that perceives itself.* For this solution to work, we must first accept that our organ of perception is a structure made of energy, and not of matter.

To use the previous analogy again, one can imagine that the pilot, who controls his remote-controlled vehicle using light waves, is himself made of light. The spirit is a pilot made of light, which controls the body using a force field, a field that can be called a "field of consciousness," "field of will," or "field of will-consciousness." This is a process similar to that of the pilot in the previous examples, who controls his vehicle using electromagnetic fields. Here, it does not matter if we do not know exactly what kind of energy or light the spirit and the field of consciousness are made of. What matters is the understanding of the operating principles, which are the same as those of other phenomena already well understood and, therefore, are not esoteric.

Until now, we have primarily relied on the universality of the laws and on the invisible to answer our questions on consciousness. To gain a clearer picture of this issue, we must also rely on the third pillar of the universalist approach: energy. Because the key to understanding consciousness is energy, light in all its known and unknown forms.

Just as we have previously divided nature into two parts, the visible and the invisible, saying that materialists are those who believe that consciousness comes from the visible part, and non-materialists are those who believe that it comes from the invisible part; we can do the same with the material and energetic parts of nature, and summarize the debate by saying that materialists are those who believe that consciousness comes from matter, whereas non-materialists are those who believe that it comes from energy. Presented in this way, the question of consciousness appears much less complex to us—there are only two choices!

When we consider that consciousness comes from energy, everything becomes clear. First, because it allows us to understand how the spirit controls the body through light, even if it must be assumed that it is a type of light that is still unknown or misunderstood by current science. But in addition, it provides us with a mechanism that can explain how the spirit perceives the world, having the ability to become "one" with the information it receives through its field of consciousness.

We live in an ocean of energy, and it is in this world of light that the process of consciousness takes place, and not in the material world, where consciousness and will only manifest themselves indirectly. In matter, we see only the *effects* of a conscious activity, which has its source in the levels of energy that are invisible to us, just as the

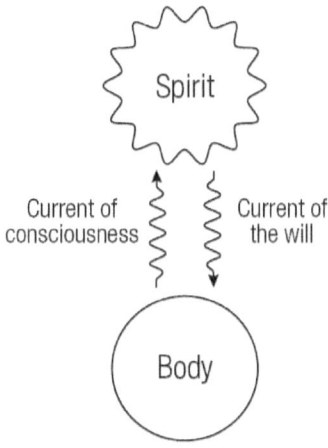

The center of consciousness and the will, the spirit, is an energetic structure. The spirit is of the same nature as the light with which it controls the body and perceives the world.

images that appear on a movie screen are just a reflection of the activity of the projector.

Therefore, within the universalist approach, we can see the different states of consciousness as different states of energy. This way of thinking also fits in the natural order. Indeed, we all know very well that nothing is ever created or destroyed in nature, that one can only *transform* what already exists—it is the law of conservation. It must be the same for consciousness: it can come only from the transformation of something that already exists, and what changes state, in the case of consciousness, is energy, not matter.

Consciousness is not a kind of strange third category, existing in addition to energy and matter, but an intrinsic property of the energetic part of the universe. As an intrinsic property of energy, consciousness is never created or destroyed; it can only change form. This does not mean that every particle of energy is conscious, but that the consciousness is always present, in a potential state, in energy. Therefore, it can be awakened and amplified according to the conditions. Consciousness is not something foreign, located outside the laws of physics, but something fundamental, inseparable from the energetic part of nature.

This is another great advantage of the universalist approach: it adds no category to those *already* known to physics! The only thing we have to accept is that consciousness comes from the energetic component of the universe, not from its material component. In other words, it is a bosonic phenomenon, not a fermionic one. Of course, this implies the existence, in energy, of structures and processes unknown to current science, but that these important realities are invisible is not surprising, since most of nature is invisible. Current science, even though it has come to understand some fundamental laws, still knows very little about the true richness of nature.

"Consciousness comes from energy." "Consciousness is light perceiving itself." "Consciousness is a state of energy." Those are different ways of summarizing the universalistic solutions to the problem of consciousness. Of course, this is only a base, but this base is extremely solid. To avoid getting lost in the details, we need to start with only an overview that focuses on the fundamentals. The natural laws are like the foundation of a building. It is only once that base is well established that one can build, stone by stone, a perfectly consistent worldview.

8.5 CONSCIOUS EXPERIENCES MADE OUTSIDE OF THE BODY

Universalism allows us to understand why no one can explain consciousness as a mechanism of the brain, it explains how the spirit can control the body through energy currents that carry information, and explains how the spirit, being made of a substance of the same nature as light, can perceive the world through it.

We will now see another category of phenomena that the universalist approach can explain: the "near-death experiences," as well as other conscious experiences made outside of the body.

The many testimonies about conscious experiences made outside of the body, abundantly found across the different cultures and eras, are very interesting, but they must not be the only reason to believe that consciousness can exist independently of the brain. Our beliefs should not be based solely on testimonials, which are always influenced by the limited perspective, the interpretations and the beliefs of the witness, and, therefore, may contain errors. There is only one perfectly solid basis in which to anchor our beliefs: *the laws of nature.*

That being said, the abundance of these testimonies of conscious experiences made outside of the body is quite consistent with universalism; it is another element that adds weight to it. Of all the testimonies of this kind, the near-death experiences are the most interesting from the point of view of contemporary science because of the conditions in which they take place. This kind of experience often occurs in an operating room, a place that is a controlled environment comparable to a laboratory. Indeed, it is close to the conditions that would need to be met in a laboratory to study conscious experiences made outside of the body: a subject linked to measuring devices that is put in a state where these devices tell us that the subject is technically dead, before restoring vital activity and collecting the testimony of what the subject experienced while he or she was considered dead. Of course, we immediately see the dangers that this kind of experience would entail, and that is why we need to be content with testimonials coming from operating rooms!

This kind of observation has some characteristics of scientific evidence, but it is not irrefutable. For example, there is the difficulty of establishing when the conscious experiences presumably made outside the body occurred, before or after cessation of brain activity, and the difficulty of measuring whether there is really a complete shutdown of the brain since the types of devices used in operating rooms cannot measure everything.

Gaps of this kind will always be emphasized by those who defend materialistic beliefs, and they are right on this point; these sources are not reliable enough to be viewed as scientific evidence. This is one of the reasons the heart of the argumentation of this book does not rely on these testimonies and these partial measures, but on the great laws, which are the pillars of science.

What matters when one approaches the question of out of body experiences with the universalist point of view is that events of this kind become normal, natural and predictable. These events cease to be anomalies, and the situation is inverted in relation to the materialist approach. These events are not strange; on the contrary, what would be strange is that events of this kind never happened! This shows us, once again, the great explanatory power of universalism, for which what are strange phenomena for the materialists become natural phenomena. Indeed, having demonstrated that the existence of the spirit is a solution that agrees with the laws of nature, it would be strange that we did not find testimony of people who have conscious experiences independent of the physical body! But since testimonies of this kind are abundant and have always accompanied humanity, it shows us, then again, that this approach is consistent with reality.

The reaction of materialists to testimonies of conscious experiences outside the body is, in turn, consistent with the attitude of some believers when faced with facts that contradict ideas dear to their eyes: *they try to diminish their importance.* For them, the question is quickly settled. They simply consider that experiences of this kind—going out of the body, seeing dead relatives or other beings, seeing a light, a tunnel, etc.—are only a weird product of our imagination, hallucinations coming from a brain in lack of oxygen, or other phenomena of this kind. This notion of hallucination is very practical for materialists because it is a versatile smokescreen, all conscious experiences that are inexplicable otherwise are placed in this very broad and vague category.

Indeed, materialists must remain vague; they cannot provide any elaborate answer to explain special states of consciousness, simply because they are unable to even explain normal states of consciousness! This is obvious, and yet, many people fall into this trap. Many are turning to neurologists asking them to "scientifically explain" extraordinary states of consciousness…forgetting that they do not even have scientific explanations for ordinary states of consciousness! If neurologists adhering to materialism are asked to explain the states of consciousness that happen when one is on the verge of death, *they can only share their beliefs with us.* They cannot transmit knowledge since neurologists admit themselves that they do not really know how consciousness works—whether or not the body is close to death does not matter! So, if our goal is to make an anthropological study of the materialistic beliefs of the 21st century, that is fine. On the other hand, if we seek real explanations, we will find nothing. It is actually the same as asking religious people to give us explanations on death; of course, they will only share their beliefs. The same goes for the materialists, they can only share with us their interpretations and not knowledge, which they admit they do not possess.

Regardless of the state of consciousness they attempt to explain, the materialists only *decide* to believe that phenomena occurring in the brain are the cause. Therefore, to accept their explanations is only to commit an act of faith, and nothing else.

The discoveries of science do not oblige us to adhere to the materialistic interpretations since all observations of neurology are also consistent with universalism, which considers the activity of the brain as an intermediate step.

Moreover, even if the brain still has low activity at the time when someone has a near-death experience, it is still necessary to question the materialistic hypothesis. People who have had near-death experiences often refer to it as one of the most important experiences of their lives, an experience that manifests itself with lucidity and an intensity that is unlike anything else they have previously experienced and can even transform them. If these experiences were really the product of a disordered brain, they should instead present themselves in incoherent chaos. But the opposite occurs: with a *decrease* in the activity of the brain, we observe an *increase* in the intensity of the conscious experience! From this observation, we can deduce that the activity of the brain *limits* the conscious experience in a manner consistent with the universalist vision. Indeed, if we consider that the role of the brain is to adapt the activity of consciousness to material reality, by slowing it down or by compressing it in some way, it is normal for consciousness to manifest itself in an extended way once released from the influence of the brain.

We have seen previously that the independent existence of consciousness is a logical necessity for solving questions that cannot otherwise be solved. Once we have understood this, the testimonies of conscious experiences outside of the body do not seem to us as anomalies, but instead, become natural and predictable phenomena. People are constantly presenting experiences of this kind as being "paranormal," whereas it is only the materialistic ideas that oblige us to qualify them this way. If we consider that consciousness is independent of the brain, these phenomena are quite normal and explainable.

The existence of these other levels of reality and the difficulty of coming into direct contact with them by means of measuring devices can be naturally explained with the law of selection; these devices simply do not have access to those levels of interaction because of their different nature. Alternatively, if a part of the human being is made of the same substances as these invisible worlds, interaction is possible. This explains why the human being has the ability to come into contact with realities that the devices of science cannot access. It is not necessary to resort to supernatural notions to explain this; the most basic principles of physics are sufficient!

From the universalist point of view, death is like any other phenomenon. It is only the brain that is no longer able to transmit the currents of consciousness and of the will because of a malfunction. This is a phenomenon comparable to what happens if a television breaks while it was transmitting a program. The second before, the device showed a lot of activity, and the second after, there is nothing. Does this mean that the source of all the information that appeared on the screen has suddenly ceased to exist? Of course not, but the transmission channel is no longer there, and since you no longer interact with the source of this activity, you are facing an *illusion* of disappearance. It is the same with death; if the brain no longer expresses conscious activity, it is because it is a broken transmission apparatus, and not because it generated this activity in the first place. It cannot be otherwise because the brain is only a material object that has no power that other material objects do not possess. It is simply a system specialized in certain processes of reception, translation and transmission of information.

We can also consider death as the separation between two substances. At the level of the principles, it is not a phenomenon different from the separation between substances as considered by chemistry. The combination of substances is possible only under specific conditions, which vary according to the case, and if these conditions are altered beyond a certain limit, this causes a separation. In the so-called "death" phenomenon, it is the body that is no longer able to provide the conditions for a lasting bond, which automatically causes a separation from the conscious part, which then displaces its activity to the invisible side of reality.

For many skeptics, to admit that consciousness and life can exist on the invisible side of reality is unacceptable from a scientific point of view. But a truly neutral and objective point of view compels us to admit that it is, on the contrary, an excellent scientific solution. Accepting the idea that consciousness and life exist in invisible forms is not adhering to supernatural beliefs; on the contrary, it means accepting *without compromise* the universality of the laws of nature. The laws must apply in the same way to the whole of reality; so, if the laws of nature allow life on the visible side, it must also be the case on the invisible side. Seeing things that way is the only way to build a worldview that is truly *self-consistent*.

It is not necessary to torture one's mind when reflecting on these questions—*basic logic is enough!* The biggest problem, when one wants to approach these questions objectively, is not the notion of invisible life, which is simple and natural; it is, rather, the mountain of false ideas that religions have invented about the invisible worlds. Because of this, religions have a great deal of responsibility for the confusion that exists over these issues.

Life after death is a very broad topics that go far beyond the scope of this book. Let us just stress that it would be a mistake to believe that we can understand everything based solely on the variety of testimonies of people who were on the verge of death or who have contact with this other reality because it is only a tiny sample, a sample that is still enough to show that the richness of life that we find in the visible world continues *uninterrupted* in the invisible worlds, in perfect conformity with the universality of the laws.

8.6 IN SUMMARY

As mentioned in Chapter 5, science progresses through conceptual unification, that is, by explaining more and more phenomena with fewer and fewer laws. It is by progressing this way that our vision of the world becomes ever more coherent.

Universalism allows us to progress toward this goal since it answers our questions using very few laws and relies only on principles already well known to science. It does not invent new laws, it simplifies our understanding of the world, explaining a lot with a handful of laws understandable by all. In what we have seen so far there are several conceptual unifications, here are the most important ones:

The brain is unconscious, like all material objects.

The center of consciousness and willpower is the spirit.

The spirit is an invisible phenomenon, like most phenomena.

The spirit is an energetic phenomenon, like most phenomena.

The interaction between the body and the spirit works according to the same laws as the other interactions.

The spirit can perceive through energy, because it is itself made of energy.

The solutions proposed by the universalist approach are simple, clear and comprehensible to all. Science already has all the knowledge needed to solve the enigma of consciousness because we do not need to wait for a "great scientific discovery" for these solutions to work. It is only necessary to *interpret differently* what science *already* knows!

Science already knows that most of nature is invisible, and it already knows that the essence of nature is energy. It is only necessary that scientists agree to put the invisible and energy at the *center* of their theory of consciousness, so that the answers appear clearly to them, answers in perfect agreement with the laws of nature. To go there, the scientific community must first abandon the materialistic beliefs that parasitize it, rejecting the preconceived ideas that prevent it from seeing the solutions it already has in its hands.

Materialists are free to continue to promote their beloved beliefs, but they cannot, without discrediting themselves, say that the universalist approach is not a good alternative from a scientific point of view, for it is a solution that possesses great explanatory power and that fits with all the known laws of science.

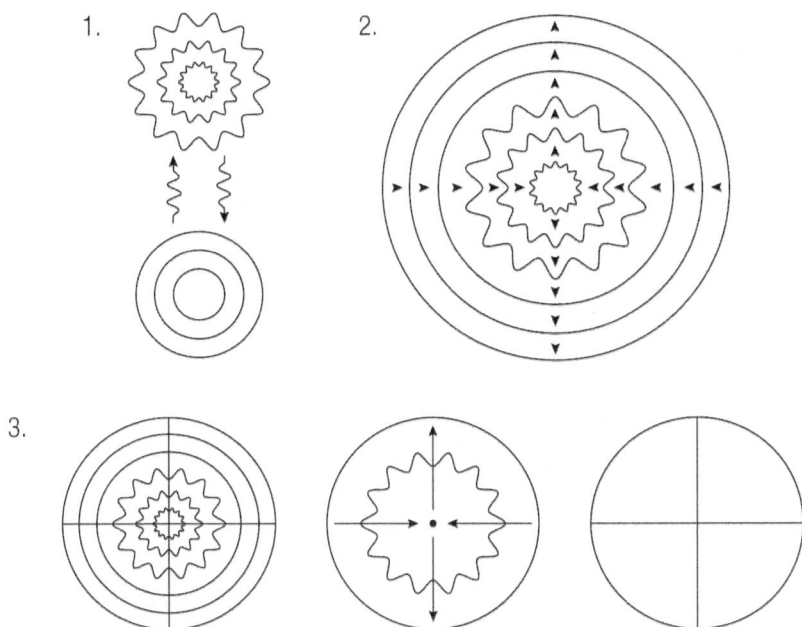

Different depictions summarizing the structure of the human being: 1. The energetic part and the material part of the human being, linked together by the currents of consciousness and the will. The energetic part and the material part of the human being each have several levels, the majority of which are invisible. 2. A depiction of the human being with the different levels inserted into each other, the current of the will goes from the center to the periphery, from the invisible to the visible, and the current of consciousness goes in the opposite direction. Each time it goes from one level to another, the current is transformed. 3. Various diagrams illustrating the same notions and representing the human being as a whole.

9. LIFE DOES NOT COME FROM MATTER

Beyond life in visible matter, there is life in invisible matter.
Beyond life in invisible matter, there is life in light.
Beyond life in light, there is Life itself.

We will now see how the three pillars of universalism (the universality of the laws, the invisible and energy) make it possible to illuminate with a new light one of the deepest questions: the origin of life. The universalist approach solves this enigma in the same way that it makes the enigma of the brain disappear. The reason materialists are not able to explain how the brain generates consciousness is that consciousness does not have its origin in the brain. It is the same for life; the reason they are not able to explain how life could emerge from matter is that life does not have its origin in matter. In both cases, these mysteries are created artificially by false interpretations based on illusions and disappear as soon as one abandons the materialistic beliefs.

As soon as we consider that life comes from the invisible side of reality, solutions in accordance with the laws of nature are available to us. Life did not appear spontaneously on Earth, it was *transmitted* from the invisible domains of nature. This is the answer proposed by the universalist approach, a solution that allows us to respect the most important law of biology: *Every life form is a reproduction.*

But before going into detail about the universalist explanations of the origin of life, let us, first of all, see why it is such a monumental problem for the materialists.

9.1 THE ORIGIN OF LIFE ACCORDING TO MATERIALISTS

Just as there is no consensus among the materialists on the process that would allow the brain to generate consciousness, there is no consensus on the process that would allow matter to engender life. Many hypotheses are confronting each other, none really stands out, and this problem appears so difficult in the eyes of the materialists that most prefer not to think about it. As they are content to believe that consciousness emerges from the brain without understanding how it is possible, most materialists are content to believe that life emerges from matter without trying to understand how. In both cases, they commit a blind act of faith, such as those found in religions, which also require us to believe without understanding.

The belief in the existence of a process that could allow matter to form organisms without the contribution of a previous life, a process sometimes called "abiogenesis," has given birth to a whole field of pseudoscience. The speculations are endless. Some believe that life has appeared thanks to molecules capable of self-replication, having acted as precursors of DNA ("genes first" hypotheses). Others believe that life appeared through chemical reactions forming a sort of primitive metabolism ("metabolism first" hypotheses). Some believe that life has appeared in pools of water. Others believe it has appeared in oceans, or in hydrothermal vents, while others believe it has appeared underground. There are endless discussions about which are the key elements to "start" life: amino acids, lipids, peptides, silicate crystals, polyphosphates, polycyclic aromatic hydrocarbons, and so on. The debate goes as far as to move in space since some advance the hypothesis that life would have appeared first on another planet, like Mars, and that it would have been brought to Earth by meteorites...

In short, the basic explanation of the materialists is always: "Life came about by accident!" And their various theories are trying to find out what kinds of blind and random processes have started life.

Being very numerous, we cannot present the materialist's hypotheses about the origin of life on Earth in detail here. But to illustrate more concretely how materialists approach this enigma, we will take, as an example, a classic experiment very often mentioned in the scientific texts dealing with the question of the origin of life: the famous Miller-Urey experiment.

This experiment was conducted in 1953 by scientists Stanley Miller and Harold Clayton Urey, and its purpose was to test whether the primitive Earth was a favorable environment for the formation of organic molecules. This experiment consists of linking two glass flasks, a flask supposed to reproduce the primitive ocean, and another supposed to reproduce the primitive atmosphere. By heating the mixture of the "ocean" flask, one produces the evaporation of the mixture, which is then found, in a gaseous state, in the "atmosphere" flask, where it is subjected to electric discharges simulating lightning. Then, the mixture is cooled so that it condenses. One then just has to let this process continue for several days and take samples to see the result.

The news of what came out of this experiment went all around the world: from simple chemical elements (e.g., water, methane, hydrogen, ammonia), the experimenters succeeded in producing more complex molecules, some of which are essential for the proper functioning of living organisms. In particular, the experiment produced certain amino acids, which are the molecules that the cells use to make proteins. In short, this experiment succeeded in making many elements that are considered "building blocks of life."

This result ignited the imagination of the materialists, who began to believe that this kind of process is the solution to explain how life could have emerged on Earth without the contribution of a previous life. Then, they began to search how these organic molecules, formed inside what they called the "primordial soup," could combine to form even more complex elements, such as primitive DNA, simple proteins or stable metabolic networks...but without significant advances since.

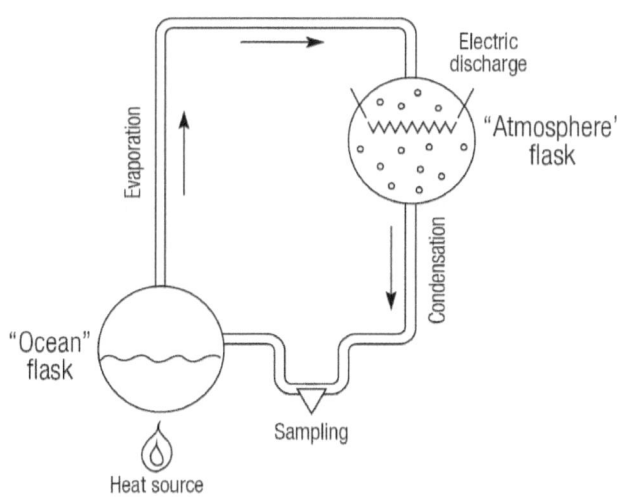

The goal of the Miller-Urey's experiment was to produce organic molecules, dubbed "bricks of life," by simulating the conditions existing on the early Earth.

In reality, with the Miller-Urey experiment, and other similar experiments, researchers do not answer the question of the origin of life—they only open the box of a huge puzzle! They have before them all kinds of pieces, building blocks that nature provides in abundance, but how could all these pieces combine by themselves to form living organisms? The mystery remains whole...

Indeed, the reason the Miller-Urey's experiment is constantly used, even more than sixty years after its realization, as "proof" that the materialist approach is realistic, is that no other experiment ever went further! In all their research, scientists have never managed to go further than the discovery of processes allowing the formation of certain building blocks of life, whether it is a chain of molecules that is presented to us as a precursor of DNA, a spherical membrane that is presented to us as a "protocell," an autocatalytic reaction that is presented to us as the beginning of metabolism, and so on. All that the materialists present to us are only pieces that can potentially enter into the making of organisms, but never anything that constitutes the beginning of an organism, even the simplest. In other words, all that scientists have discovered can no more be seen as a "primitive organism," than a brick can be seen as a "primitive house"!

How could these primitive organisms, which the materialists conceive in the marvelous laboratory of their imagination, have been formed from the building blocks of life, going against the pressure of an environment that constantly threatened to destroy them? This question is a huge headache for the materialists because, in their vision, the organizing pressure of life was initially absent, pressure that is necessary to resist the environment and gradually produce evolution by natural selection. For them, in the beginning, there was no organizing pressure, only random activity. That is, molecules that collide, combine, and separate randomly. It is this blind activity that must produce, by the succession of the most fantastic chance events, stable systems capable of reproducing, systems that can withstand the pitiless pressure of the environment, and that long enough to be able to produce many descendants, descendants that must have enough variety to be able to adapt and ultimately to evolve thanks to natural selection... How can we not have headaches when thinking about all the incredible feats that chance had to accomplish to form the first organisms?

To put things in perspective, let us imagine that a group of researchers managed to produce in a laboratory, from the building blocks of life mentioned earlier, a microscopic system that would be a kind of molecular machine. Then, imagine that this tiny machine is able to repair itself automatically if it is constantly provided with materials that it can assimilate, and, on top of that, this molecular machine is able to use these materials to make copies of itself! Researchers who would realize such a feat would surely be considered among the greatest geniuses. Yet, this feat, which the best researchers have never managed to achieve in the controlled conditions of a laboratory, materialists believe *chance*, under the *random* conditions of the environment, is able to do it!

Materialists themselves find the idea of a machine forming by chance absurd. Yet, when they think about the formation of the first organisms, which are only molecular machines, they install an artificial division in their thoughts, believing that in this special case it is possible. To believe that chance can achieve such feats is to commit a gigantic leap of faith. In reality, the intensity of the blind faith that materialists have toward chance is quite comparable to that found among the most fanatical religious people! How can rational beings believe that chance can have such power when everything in science tells us that it is the exact *opposite* of what chance can produce?

Do we really need to explain why machines, organisms or any other object presenting the same kind of complexity, cannot be formed by chance? Everyone understands intuitively that this is impossible, even the materialists, who make an exception only in the case of the first organisms, invoking miracles produced by chance because it is necessary to preserve their cherished beliefs.

Even if this idea of organisms forming spontaneously is widespread in scientific circles, it is not proof that it is rational; it is only another proof of the great power of beliefs! A false belief can cause a normally sensible person to reason in an illogical way when facing a particular question because, in this area, he or she has decided to adhere to beliefs disconnected from reality. This is the case with materialists, who believe that life can be summed up as interactions between material elements, and who, therefore, imagine that life should be able to emerge only from these interactions. Therefore, according to them, the first organisms had to emerge from interactions which, in the beginning, were purely random. In approaching the problem in this way, they must reason *upside down* because such an organization is exactly the opposite of what chance can produce!

Materialists themselves admit that what they believe does not enter the natural order of things, since their mythology contains the idea that the chain of events that led to the emergence of life is "highly improbable"—another expression that is only smoke and mirrors since this chain of events is not "highly improbable," but *absolutely impossible*. The fact that objects as complex as organisms cannot form spontaneously comes from the most basic physics, and everyone already understands it intuitively. But, to give a complete portrait of the subject, let us see why the laws of nature forbid it completely.

9.2 THE LAW OF EQUILIBRIUM

Physics studies the behaviors of energy and matter within space and time. We are all impressed by these large blackboards that physicists fill with mathematical formulas, which create the illusion that nature is something very difficult to understand. But, when we go beyond the mathematical language, we quickly realize that the behaviors studied by physicists obey a set of very simple laws, which all can understand even without knowing how to read or count.

We have already encountered some of these laws, such as the laws of interactions, selection and retroaction. These are all laws that express themselves through the formulas used by physicists, but which, in this book, have been presented to be understood intuitively, and not mathematically.

To understand why complex objects, such as machines or organisms, cannot be formed by chance, we must add another law to this list. This law is the law of equilibrium: *Everything tends toward equilibrium.*

Every movement tends toward a point of equilibrium, following the path that deviates the least from equilibrium. That is, the shortest possible path, the one that involves the smallest expenditure of energy, which minimizes the action. A path that, under ideal conditions, will always be a straight line. This is what the physicists sometimes call the "principle of least action." Physicists have many ways to express the law of equilibrium through mathematics, but the results can be summed up in the simple fact that all movements tend toward equilibrium, following a path that is as straight and short as possible, and a good deal of the formulas of physics only serve to find what this path is.

To understand why this principle of physics forbids the formation of organisms through the miraculous power of chance, it must be applied to another important area studied by science: mixtures.

Remember that the materialists consider that life appeared in what they often call the primordial soup, which is a mixture of organic molecules, also called "building blocks of life." The exact composition of this mixture and the place where it was formed vary according to the hypotheses, but such details are of no importance here since the consequences of the law of equilibrium are the same in all cases.

Again, the consequences of this law are very simple: *Every mixture tends toward equilibrium.* Every mixture tends to adopt the most stable state possible, which is ideally a uniform configuration, in which energy is distributed evenly.

Energy always tries to spread evenly. It is for this reason that, when a drop falls on the surface of a body of water, the energy spreads in concentric waves. It is also for this reason that light spreads evenly from a light bulb or the flame of a candle. It is the same principle that is at work when the energy spreads inside a mixture; it will always seek to distribute itself equally, which has the effect of pushing the elements in the mixture to distribute themselves evenly.

Again, these consequences of the law of equilibrium, which are outlined here, are already intuitively known to all. For example, when we mix ingredients while cooking, we know naturally that the elements we combine will be evenly distributed in the mix. We do not need to make an effort to consciously distribute the ingredients evenly in our omelets, the laws of nature do this in our place. We just need to transmit energy to the mixture by stirring it!

If we leave a window open in winter, we know that it will get colder and colder inside. This is because the air itself is a mixture that tends to become uniform, a rule that also applies to temperature, and by opening the window, we offer the possibility for the indoor and outdoor air temperature to become the same. The same thing happens when we put clothes out to dry on a clothesline since the humidity level between our clothes and the ambient air also tends to become uniform.

This law is also the reason we constantly have to clean our homes. Indeed, from the point of view of the laws of nature, a house is only a mixture of different objects, which, like all mixtures, tends toward uniformity. That is why, when we stop doing the housework, all the objects tend to spread more and more evenly. Clothes that hang around everywhere, dishes on every corner of the kitchen counter, children's toys that watch our every step, dust that gradually invades the space—even the cat tends to spread evenly, leaving its hair everywhere!

These are all consequences of the law of equilibrium, the same law that forces the atoms to adopt a uniform configuration inside crystals, as much as it distributes the galaxies uniformly in the universe... The great laws of physics are the same laws that dictate our everyday life because there are no other laws!

The reason the materialist theories of the emergence of life do not work is perfectly simple, and everyone already knows it intuitively. The reason is this: It is impossible for complex systems to emerge from a mixture by random processes because it goes against the trend toward uniformity that *all* mixtures must obey!

The materialists have been racking their brains for generations, trying to understand how the building blocks of life could have combined within the primordial mixture. They reflect about all kinds of constraints that could have forced these elements to combine, but no matter what the constraints are, the laws of nature only push the elements to be evenly distributed within these constraints—and that's it! This is why the solutions imagined by the materialists always fail the test of reality. Nature only distributes the elements in the most equal way possible, without ever producing the magical combinations that materialists hope for, because if we leave a mixture in the hands of randomness, it is impossible for anything else to happen! Combining molecules to form complex systems requires moving far away from the uniform state, which is why it is impossible for this to happen as long as it depends on chance. It is like hoping that a rock that we let go will go up rather than down, or for a river to suddenly start to travel upstream! All these phenomena require an interruption of the laws of nature, something that has never happened and will never happen.

The law of equilibrium alone suffices to refute all the materialist theories of the emergence of life. The fact that this law is not respected is a huge logical flaw, which is enough to raze the building of materialism to the ground. Materialists are literally making chance their god, believing that it has the miraculous power to bring life to inert matter. Yet, if they opened their eyes, they would see for themselves that what they require of chance, in the case of the origin of life, is exactly the opposite of what chance can produce.

The law of probabilities is very clear: Everything that depends on chance tends toward its most probable state. In the case of any mixture, the most probable state is a uniform one, in which the elements are distributed as evenly as possible. This phenomenon is well known to physicists, who use the concept of "entropy" to talk about this inevitable tendency toward disorder. Entropy measures the level of energy degradation of a system, and one of the most important laws of physics states that the entropy of a system tends to increase over time until an equilibrium is reached; it is the second law of thermodynamics.

The law of entropy stipulates that the energy of a system left to itself can only become less and less available over time, which is the same as saying that energy tends to be distributed more and more evenly over time, as mentioned earlier. The tendency of entropy to increase is just another consequence of the law of equilibrium. It is always the same behavior of energy, which pushes everything toward equilibrium, toward the most uniform state possible. By acting within a system, this tendency has the effect of gradually decomposing the elements if an organizing pressure is not constantly in action to prevent it. This is why entropy is often presented as a tendency toward disorder, even if it is more precise to speak of a tendency toward uniformity and equilibrium.

The law of entropy only repeats, with scholarly words, what everyone already knows: what is not constantly being maintained tends to break down with time! This is what happens when an organism dies. When the flow of information that comes from the expression of genes ceases to do its organizing work, the organism finds itself completely at the mercy of the fluctuations of nature, which rapidly begin to break down the organism, to disorganize it, to blend it with its environment. Because the only thing that can prevent this tendency toward uniformity is a constant organizing pressure, obeying precise information, as it is the case when the genes express themselves by forming elements that serve for the maintenance of cells. *Precise* information is the *only* thing that can counteract the disorganizing effect of chance, so life could *only* have appeared on Earth through the transmission of well-defined information, that is to say, which leaves little room for chance, imprecision, noise. All other explanations are in contradiction to the laws of nature.

Materialistic scientists are at war with the most firmly established laws of physics, chemistry and biology. They try to combine all sorts of little tricks with the hope of reversing the inevitable effects of these great laws. It is like trying to topple a mountain by throwing little stones at it!

Everywhere in nature is inscribed this great law: *Every life form is a reproduction.* Only life can give birth to life, what is inanimate cannot. The belief that in a distant past, inert elements accidentally combined to "start life" is a pure product of the imagination. Materialism is centered on one of the strangest superstitions: *the belief that what is inanimate can give life.* Materialists like to present themselves as "defenders of reason," saying that their philosophy can free us of superstition when, in fact, they have only invented superstitions of a new kind.

9.3 MOST OF LIFE IS INVISIBLE

At first sight, it may seem incredibly pretentious to propose, in a single book, answers to the "mystery of the origin of life" in addition to solutions to the "mystery of consciousness." Indeed, those who argue that it is possible to find simple answers to these questions are automatically discredited by most, so strong is the belief that these are the most difficult subjects there is.

But, in reality, these so-called mysteries are not complex problems. This impression is born only because the discussions on these subjects, in scientific circles, are parasitized with all kinds of false beliefs based on deceptive appearances. Of course, people who take seriously the materialist theories currently in vogue can only end up in a state of confusion, but this is only because they have given value to ideas disconnected from reality. This confusion does not come from nature, the rules of which are perfectly clear.

This is always how science has progressed, phenomena previously considered as impenetrable mysteries have subsequently proved to have logical explanations rooted in simple natural laws. Consciousness and life are natural phenomena, and, like everything else, they must have logical explanations in accordance with the laws. The goal of universalism is to deepen these natural explanations, recognizing, first of all, that to see these questions clearly, we must put the invisible and energy at the center of our theories of consciousness and life, and not at the periphery, as is the case in materialistic thought.

The reason it is possible to discuss, in the same work, as much of the enigma of life as of consciousness, is that the answers are fundamentally the same! In both cases, the answers are in the universality of the laws, the invisible and energy. Let us now see how, following reasoning similar to those we have used for consciousness, we can see more clearly about what many consider to be the greatest of all mysteries: the origin of life.

The first key we will use to solve the question of the origin of life is the same one we used in the case of consciousness: *Most of nature is invisible.*

Throughout this book, we have emphasized this fact. The universe is filled with unknown substances that we cannot perceive directly with our senses or instruments, but the existence of which can be understood using logical deductions based on natural laws. From there, we have seen that it is quite logical, as a part of the solution to the enigma of consciousness, to consider that most of the human being is also invisible. To say that most of the human being is invisible is, in fact, the same thing as to say that the human being is part of the natural order, where the invisible is always much more important than the visible.

Similarly, to progress on the question of life, we must first accept that most of life is invisible. So, as we have previously represented the human being in two parts, visible and invisible, we will begin our reflection on life by conceiving that it also exists in these two broad categories.

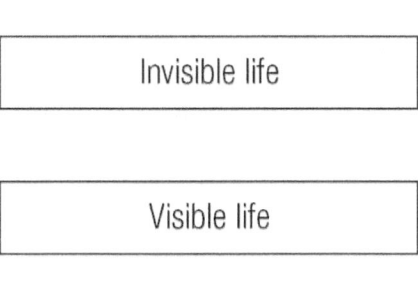

In this book, when we speak of invisible life, we only speak of a life that exists in the invisible substances that permeate the universe, a life that is not fundamentally different from visible life. Here is a simple definition of invisible life, as it is considered here: *Invisible life is a life that functions according to the same laws as visible life, but that exists in substances invisible to us.*

The principles that define invisible life must be the same as those that define visible life, otherwise it is not "life!" That is to say, this invisible life must be made up of a great variety of interdependent species, able to reproduce and evolve, just like visible life. As for knowing exactly what forms this invisible life can take, this is an inexhaustible question since it must exist in an almost infinite variety of forms, like visible life! The question of the forms of this invisible life does not matter here. Only the principles, the outlines, are important since it is only by remaining at the level of the principles that we can see clearly. No matter what forms the invisible life may take, it is governed by the same laws as visible life, since these are universal.

Through all our reasoning, the basis of our reflection must always remain the universality of the laws of nature. Here is another example of reasoning based on the universality of the laws in favor of the existence of invisible life.

In recent decades, astronomers have been able to observe planets orbiting stars other than the Sun, whereas these exoplanets were previously invisible to our telescopes. Yet, long before these observations were made, scientists already considered the existence of planets around other stars as self-evident. Once again, their hypotheses were based on the universality of the laws; the fact that the Sun has planets is an indirect proof that most of the other stars must have planets too since all the stars are subject to the same laws. It is also by this same principle that most scientists believe that life must exist on planets other than Earth, even if this life has not yet been observed, because the universal laws that make life possible on Earth must make it possible on planets similar to Earth.

We must follow the same reasoning when thinking about the existence of invisible life. If life is possible in visible substances, it must also be possible in invisible substances since the same laws are at work in both fields. To believe the opposite *is to give a special place to the visible,* a position that goes against the universality of the laws and the principle of relativity, as we saw in Chapter 6.

It is like believing that the Sun is the only star to have planets or to believe that the Earth is the only planet to house life. That is wrong because it gives our star or our planet a special place, which is prohibited by the principle of universality. Likewise, from the point of view of the laws of nature, *the visible is no more important than the invisible.* The visible domain is not special, it is only a level of existence, among many others! We must always remember that nothing is visible or invisible in itself, it is only a question of point of view. The "visible" is only the small number of substances with which our senses or instruments are able to interact. If our senses or instruments were of a different nature, what is visible to us would become invisible, and the invisible would become visible, appearing to us just as full of structures and phenomena as the small fraction of nature that we can normally perceive.

The notions of "visible" and "invisible" do not exist for nature. There exists only a great variety of substances, all subject to the *same* laws. If these laws permit life in substances known to present-day science, this is indirect proof that life must also be possible in the unknown substances, which are said to be "invisible."

Humanity has understood that the Earth is only one planet among many others; it must also understand that the visible is only one level of existence, among many others. From a scientific point of view, the proof of the existence of invisible life is simply the fact that the visible cannot have a special place in relation to the rest of reality, which would be the case if it was the only domain where life is possible. In short, the proof of the existence of invisible life is *the universality of the laws of nature.*

Materialists are quite right to be suspicious of all that is said about life in the invisible realms. But they themselves cease to be rational when they use these inventions as a pretext to reject even the very existence of this invisible life. Again, what is irrational is not believing in the existence of invisible realities, but to conceive of these domains as outside of natural laws!

We must not believe in the existence of an invisible life only to conform to a religious or spiritual philosophy, but because it is a natural and logical notion, indispensable to have a correct vision of reality! Invisible life is a perfectly rational notion; it is even completely absurd to reject it since it allows us to solve some of the greatest mysteries of science. Many of these so-called mysteries were created from scratch when scientists began to deny the existence of invisible life, seeing it only as a vulgar superstition; thus, they have deprived themselves of an indispensable element to understand nature, which has been sorely missed since.

Nowadays, the subject of invisible life is like an impenetrable jungle, where the true and the false intermingle in a nameless confusion. To get out of this jungle, it is necessary that our thoughts always follow the strongest guiding principles there are—the laws of nature. This is the only thing that can really help anyone who wants to understand this subject in a rational way.

It is these laws that make it possible to explain why human beings can come into contact with realities inaccessible to the instruments of science. This is explained naturally as soon as one accepts that the human being has invisible parts, like those worlds harboring an invisible life. This gives the human being the ability to interact with these other levels of existence, which the instruments of science cannot do, lacking the necessary elements to serve as a bridge.

Among other things, this interpretation gives us an explanation of what our inner life actually is. Instead of being interpreted as an inexplicable by-product of brain activity, it appears to us as a life just as real as visible life, which we perceive with *senses* that are other than those of our physical bodies. Materialists see these as weird ideas, whereas, on the contrary, it is the most natural way of explaining what our inner life is. Universalism considers the inner life as the product of perceptions functioning in the same way as other perceptions. In other words, it considers that *all* perceptions function in the same way, both those of our inner life and those of our external life—an interpretation that, once again, agrees with the universality of the laws!

From this point of view, it is wrong to consider the ability to come into contact with other levels of existence as being possible only for a handful of exceptional beings. We are all constantly interacting with those domains, but normally it is the perceptions from the physical world that are in the foreground. What happens, in cases where people have experiences where they fully perceive what is normally in the realm of the invisible is that this usual order is reversed; perceptions of the domain of the inner life came to the foreground while those of the physical world have tilted to the background. In other words, the connections are changed, allowing the inner life to reach consciousness more powerfully.

This can happen for a lot of reasons. For example, it may be because this person is naturally more sensitive than average to the invisible realm; this can also occur when the brain is not in its usual state, such as when it is in a state close to sleep, or under the influence of drugs; it can also happen when the body is close to death.

In the current materialistic interpretation, it is said that this person is "hallucinating." We are offered this explanation as if it is the only one that agrees with science, when it is false. To say that these perceptions are possible because the human being has other senses aside from those of the physical body is also a scientifically acceptable explanation. It suffices to consider that the human being possesses invisible parts endowed with their own senses to obtain an explanation that fits entirely with the natural laws.

All this is explained by elementary physics because the law of selection is at work here. All interactions are selective; two particles that do not obey the same kinds of interactions can coexist without ever interacting, as if one was nonexistent for the other, whereas the particles of the same kinds can form complex interaction networks. Therefore, structures formed by one kind of particle can exist in the same space as structures formed by another genre because their exchange networks are different and do not connect with each other.

It is this principle that allows worlds, made of different substances, to coexist without interfering with each other. These worlds exist in parallel, as layers or levels, in a way comparable to the radio frequency range, which allows different broadcasts to exist in the same space without interfering. On each of these levels, life can take different forms, life forms with which the human being can potentially interact, possessing some of these invisible substances.

The fact that there are natural ways to explain the existence of invisible worlds, and to explain why humans can perceive what the instruments of science cannot, is enough to give these ideas scientific value. Simply because these ideas allow us to *explain* many phenomena, that they provide *concrete* answers to many questions, relying *solely* on laws well tested by the scientific method.

We only need to open our eyes to see, everywhere around us, the incredible diversity of life that extends in all directions; then, it suffices to accept that this same diversity exists *also* in the invisible worlds, for the notion of invisible life to appear to us as the most natural thing.

Nature is much richer than we think—much, MUCH, **MUCH** richer! The universalist approach does not give any importance to appearances; it traverses them like ghosts, to build directly on the *base* of reality: the laws of nature. These laws give birth everywhere to an inconceivable variety of phenomena, and it must be the same in the parts of nature that are invisible to us.

What must be avoided at all costs are conceptions that are too limited. All that is encompassed here under the expression "invisible life" actually contains very many levels, within which there are different kingdoms of nature, such as the mineral, plant and animal kingdoms that we know on the visible levels. It is the knowledge of these other reigns, very fragmentary and caricatured, that has reached us through religious accounts, myths and legends. It is also the same for the invisible part of the human being, which contains several parts made of different substances, interacting organically. Everything about invisible life and the invisible human being is as rich as visible life and the visible human being. And all this complexity, seemingly inextricable, is animated by the same laws, which are perfectly simple!

To see clearly, we must learn to focus only on the laws, the main lines, and not on the infinite variety of details. It is at the level of the laws that our reflections must take place to avoid getting lost in the labyrinth that surrounds the questions of the true nature of life and the human being, a labyrinth that the mystifiers, religious or materialist, have zealously maintained for millennia. To rise above this artificial mist, it suffices to accept this one truth: *The answers to the big questions are simple because the answers are in the laws of nature, which are simple!*

9.4 THE LAW OF REPRODUCTION

All understand intuitively how the existence of life in the invisible substances that surround us can explain certain phenomena that are usually classed in the domain of the paranormal or the supernatural, whether we talk about the existence of what is called "the beyond," contact with entities of all kinds found in religious narratives or folklore, and so on. On the other hand, what is less clear to many is how the existence of an invisible life makes it possible to solve certain scientific mysteries.

One of the enigmas that the existence of invisible life can illuminate with a new light is that of the origin of life on Earth, simply because it allows us to understand that life did not appear in matter by chance, but that it was rather *transmitted* from the invisible realms, which allows us to respect the fundamental law of biology: *Every life form is the reproduction of another life.*

This law, which may be called the "law of reproduction," is the central pillar of biology. All life is transmitted from another life, never an exception to this rule has ever been observed, in any form whatsoever—it is one of the most firmly established laws of all sciences!

The belief that it is possible for life to emerge from matter spontaneously is certainly widespread in scientific circles, but this idea is an invention of the materialists, it does not come from science itself. To qualify as "scientific," this idea would need to be confirmed by experiments, a test that it has always failed for the reason mentioned previously: the law of equilibrium completely refutes it. The solutions that materialists conceive of only work in the marvelous laboratory of their imagination, where the laws of nature can be suspended. All the observations that have been made in *reality* refute the materialist's conception of the appearance of life, in a way that is perfectly clear, but that materialists refuse to see.

The language of science, the language of reality, is the logic of the laws of nature. To understand the laws is to understand reality since everything that is real owes its existence only to the laws and nothing else. Every phenomenon is born by the laws, lives by the laws and dies by the laws. No phenomenon can deviate from the paths that the laws allow, even by the thickness of an atom.

We must constantly insist on this point since *all* the confusion that exists around the so-called mysteries of life comes from the fact that we too often neglect the importance of the laws, suspend them in our imagination to accommodate our beliefs, an error that both materialists and religious people commit.

The universalist approach is centered on the laws and nothing else. The law of reproduction is a fundamental law of nature, it is universal and must apply to all life forms—even the first forms of earthly life! To accept that the law of reproduction is universal makes it possible to see the question of the origin of life on Earth in another light. Because life is always transmitted, the first terrestrial life forms, and by extension the first forms of material life, could not have appeared on their own; *they must have been transmitted from a previous life.* In other words, the law of reproduction means that life must necessarily have existed in invisible forms *before* being transmitted in matter in visible forms.

Of course, some will say that it only displaces the question, since one has to wonder then where this invisible life comes from. This question also finds its answer in the law of reproduction, but to advance only one step at a time, it will be dealt with later. For now, let us focus on the emergence of life on Earth, to understand how it is possible to solve this puzzle in a natural way, considering that the emergence of visible life comes from a transmission from invisible life.

All life forms are reproductions; to be more precise, every life form comes from a *transmission of information from a model.* As everyone knows, this transmission is conducted by means of reproductive cells, seeds, which relay information between models and reproductions. Whether these seeds are eggs, spermatozoa, spores, pollen or a unicellular organism that can be divided into a copy of itself, the process is always the same: there are initially models, these models transmit information in the form of codes, and this information is then decoded to form copies. This process is summarized in the following figure.

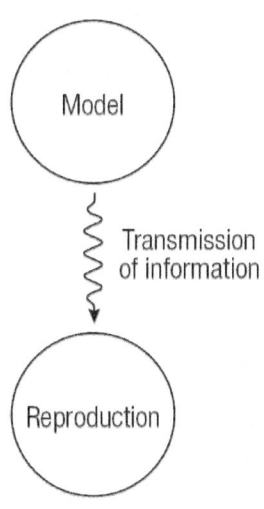

This process that is found in the field of biology also applies in many other areas. For example, when we communicate with someone using a phone, the same process occurs. The phone records our words, then encodes this information as signals that are sent to the other person's phone, and this device then decodes that information to form a *reproduction* of our words by vibrating its speaker.

It is the same process that is at work when we watch television, what we see on our screens is only a reproduction of the original information from the studio that produced the program. This process is also the basis of the industry, which, on its assembly lines, only produces copies from original plans. It is also found at work in photocopiers, where the original information is reproduced using information transmitted by light. This process can also be found in art, as, for example, when a painter reproduces a model using the information he captures through his eyes, or when a musician reproduces a melody thanks to the information contained in his scores.

As the last example, let us also mention the process by which we perceive the world. Indeed, we must not forget that our senses also do the same thing, they encode information about the world around us before transmitting it to the brain in the form of signals through the nerves. The brain then uses this information to form a reproduction of our environment, an image, and it is this partial image of the world that we perceive.

The law of reproduction is at work everywhere, down to the fundamental level of reality, in wave phenomena. This, because waves propagate only by *successive reproductions*. A wave is a pattern that moves by being reproduced from one place to another, and since physics tells us that, from a certain point of view, everything is made of waves, it means that the law of reproduction is fundamental.

In the field of biology, this law can be worded as follows: "Every life form is the reproduction of another life." This law can also be considered in a general form, to become: "Every form is the reproduction of a model." It is this law, perfectly simple and natural, that is the key to the mystery of the origin of life.

To be able to use this law to explain the emergence of life on Earth, we only need to accept one point: The models at the origin of life on Earth are invisible. As soon as we accept this point, everything becomes clearer since it allows us to consider the appearance of life on Earth as a reproduction process, similar in all respects to other reproduction processes! Here is another version of the previous diagram, summarizing this solution (next page).

This schema alone summarizes the universalist solution to the problem of the emergence of life on Earth. The *only* difference is that the models are invisible life forms; for the rest, it is exactly the same process we continually see happening around us! Nothing esoteric, nothing magical, it is only a question of recognizing the true importance of this great principle: the law of reproduction.

Science makes it possible to conceive the existence of life in invisible substances, whether or not we understand in what form. On the other hand, it does not make it possible to conceive a life form that is not a reproduction *because it contradicts all the observations that have been made on the functioning of life!*

In all their normal thinking about the living, materialists consider the law of reproduction to be an unshakable pillar of

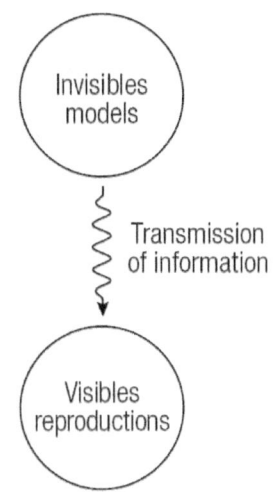

science, except in the field of the origin of life, where they are ready to throw this pillar to the ground. It is this gesture that has created the thick mist of artificial confusion in which the materialists are lost, a mist that emanates from their own intellect, and which they pretentiously call "the great mystery of the origin of life." All this confusion comes from the fact that they do not truly respect the law of reproduction, that they do not consider it a universal law.

Our ignorance of the invisible reality is great, but we know at least one thing with certainty: The laws must apply in the same way in this area because of the principle of universality. It is the same for invisible life, we do not ignore everything of this life since we know that it must respect the same laws as visible life. Invisible life remains fundamentally the same as visible life, no matter in what forms it exists. The primary difference is only that it exists in substances poorly understood by current science; in other words, it exists in *habitats* different from those known to us. For all the rest, it must be the same thing; that is, this invisible life must consist of an inconceivable variety of species, forming ecosystems of incredible complexity, like visible life.

The notion of invisible life is not strange since it fits perfectly with the laws of nature; only, in our time, people are used to thinking that the notion of invisible life is not "scientific." Even though the majority of humanity believes that an invisible life must exist in one form or another, this invisible life is generally placed in a mental domain where science and reason have no place. This conception of invisible life is completely false and must be changed. Such an artificial division can only create confusion since then it becomes impossible to put together our beliefs regarding life in a single worldview that is coherent and elegant.

To believe in the existence of invisible life is not at all irrational. It is enough to study the history of science to see how often invisible solutions have been proposed in the past by people who scientists today consider models!

For example, early in the development of modern medicine, there were all sorts of theories attempting to explain how contagious diseases could spread. Some researchers dared to suggest that these diseases were actually carried by tiny organisms, invisible to the naked eye, and, of course, this idea was mocked by many "experts" of the time. Some doctors even thought it was ridiculous to wash one's hands before carrying out an operation to avoid the risk of propagating pathogenic organisms. How many unnecessary illnesses, and even deaths, were provoked by this stupid attitude?

In physics, too, there are many examples of the power of invisible solutions. For example, there is the Higgs boson, the existence of which was predicted decades before its observation, and which was, therefore, an invisible particle for our instruments for a long time. There are dark matter and dark energy, which are believed to form about 95% of the universe, substances that have not yet been observed directly, but many astrophysical observations of which support the existence. As another example, there is Michael Faraday, at the beginning of the 19th century, who proposed that the wires in which an electric current circulated were surrounded by invisible force fields. Nowadays, the existence of electromagnetic fields is a banality, but at the time, this idea drew him mockery from some of his colleagues, as is often the case when a researcher dares to assert that he or she believes in the existence of invisible things. Today, anyone who is interested in the history of science knows the name of Faraday, but who remembers the names of these mockers?

Historically, the scientific community has a strange love-hate relationship with the invisible. It constantly disparages the recourse to invisible solutions, seeing it as magical thinking, while recognizing that many great discoveries first existed in the form of invisible solutions!

Among the many examples of this, let us see another story that took place around the year 1930 in the field of physics. At that time, physicists had difficulty with a particular phenomenon called "beta decay." Without going into detail, beta decay occurs when certain atomic nuclei spontaneously transform, emitting particles. When this process was discovered, these were named "beta particles," but later it was realized that they were actually electrons and positrons.

Physicists studying this process at the beginning of the 20th century were faced with a puzzle since the measurements of the energy emitted by beta decay did not fit with what their formulas predicted, they were systematically lower. Around this problem, two schools of thought were formed. The first school, centered around the physicist Niels Bohr, considered that these measures meant that the law of conservation of energy was not an exact law and that at the level of the particles, it could be allowed to lose a tiny bit of energy during transformations. The other school, centered around the physicist Wolfgang Pauli, rather thought that these measurements pointed to the existence of an invisible particle, the neutrino, and that the missing energy was transported by this particle never before observed.

The debate lasted for some time, and it was still raging at a conference that took place between physicists of the time, in 1932, in Copenhagen, Denmark. On the sidelines of this congress, some participants put on a play, which was a parody of Goethe's famous *Faust*. In this play, a man, Faust, signs a pact with the devil, Mephistopheles. In this parody, it was even decided to make Mephistopheles a caricature of Pauli, who, at a certain point in the play, tries to convince the poor Faust of the validity of the idea of the neutrino!

This parody sums up very well the attitude of the scientific community toward invisible solutions. It is not surprising, from a symbolic point of view, that it is the "devil" who supports the idea of the neutrino since invisible solutions are generally considered "evil" by scientists; in other words, invisible solutions are seen as a kind of scientific sin that must be avoided. Of course, scientists do not use these religious terms when speaking of these issues, but the mental processes are similar...

As for the neutrino, the formulas of Pauli, and his colleague Enrico Fermi, were convincing enough to gradually rally the scientific community, and the question was definitely settled with the detection of the particle in 1956. This was the end of a debate that became a classic case in the history of physics.

In summary, Bohr's mistake was to believe that the measurements showed him all of what was to know about beta decay, and this error pushed him to question one of the most important laws of physics: the law of conservation of energy. Pauli, for his part, continued to believe that the law of conservation must be respected in all cases and, therefore, that part of the energy must be contained in a particle that escaped measurements. Thinking in this way allowed him to discover the neutrino, which is one of the most abundant particles in the universe. At every moment, billions of these particles pass through every square inch of our body, regardless of whether or not we believe in their existence!

This is how science really works, and this is an example of the great discoveries that can be made when we trust the laws rather than the partial measures of our instruments. It goes without saying that since this happened, there is almost no physicist who dares to question the law of conservation of energy!

Just as the law of conservation is a pillar of physics, the law of reproduction is a pillar of biology. Researchers who wonder about the origin of life would benefit from learning from the history of science to see what happens when fundamental laws are neglected, only to fit with imperfect measures. This is what materialists who reject the existence of invisible life do, only because it escapes our instruments. And, to compensate, they must elaborate all kinds of convoluted theories to explain how life could emerge from matter spontaneously, theories that systematically disregard the law of reproduction, the law of equilibrium, as well as many other laws!

Skeptics will surely say that to propose the existence of new invisible particles, like the neutrino, is much less speculative than to suggest the existence of invisible life forms and, therefore, that we cannot really compare these two domains. But, in reality, the universalist approach offers nothing very speculative or new. Science already knows that reality is essentially invisible; we only suppose that this invisible part of nature also contains life, like the visible part, and we do not make this supposition by speculating in a whimsical way, but by using the universality of the laws as a support, a principle that is the main pillar of science. Also, to consider the notion of invisible life as something new is absurd since this notion has accompanied humanity since the dawn of time. From a historical point of view, *it is materialism that is an anomaly*, not the belief in the existence of invisible life. All that the universalist approach does is explain why the existence of an invisible life is natural and logical, contrary to what the materialists say, and explain how this invisible life solves the problem of the origin of life on Earth.

Science does not require that we avoid invisible solutions, but only we must have good reasons to use them, and wanting to respect the basic laws of nature is a great reason to believe in the existence of invisible life. To emphasize this fact even more clearly, let us now look at one last example of the importance of invisible solutions in the history of science, an example directly related to the question of the origin of life: the old beliefs in the spontaneous generation of living organisms.

As we have seen, the idea of spontaneous generation is the belief that it is possible for organisms to form spontaneously when the right conditions are met. It is an ancient belief, traces of it are found among the philosophers of antiquity, and it was widespread before the advent of modern science, as well among the uneducated, as among the learned.

Some version of this belief concerned especially the smallest organisms. For example, it was believed that maggots, the larvae of flies, form spontaneously in rotten meat; that fleas could be born of dust; that mice could appear in piles of straws; that some mollusks might emerge from the mud by themselves; and later, with the advent of the microscope, this belief was also applied to microorganisms.

Today, we smile at these ideas, but we must not forget that they were maintained by the greatest intellectuals of the time, and that, if we could go back to the past, it would be us who would be ridiculed to doubt the reality of these phenomena that most considered obvious!

Those kinds of spontaneous generation were increasingly questioned with the development of modern science. For example, the Italian Francesco Redi, in the 17th century, proved that fly larva did not spontaneously originate from meat, showing that if a piece of meat was protected from flies, it did not become covered with maggots, unlike the ones that were exposed. This common misconception then found a commonplace explanation: maggots do not come from the spontaneous transformation of meat, but from tiny eggs that flies lay in meat.

With those kinds of discoveries at the time, even though the idea of spontaneous generation fell into disuse for organisms visible to the naked eye, many still believed that it was possible for microorganisms, the existence of which had just been discovered thanks to the invention of the microscope. It was not until a series of experiments by Louis Pasteur, in the middle of the 19th century, that this belief was also proved false. Pasteur refuted the spontaneous generation of microorganisms in much the same way that Redi had done for maggots, that is, by protecting a substance from sources of contamination. He sterilized liquids by boiling them and then showed that the sterilized liquids that were protected from dust floating in the air did not become populated with microorganisms again, unlike those exposed to them. These experiments showed that it was the microorganisms traveling on the dust that came to colonize the liquids.

These discoveries, which validated the law of reproduction each time, allowed scientists to get rid of an artificial division and develop a more unified vision of life. These experiments showed that there are not two categories of life, that is, "life that is a reproduction" and "life that emerges spontaneously," but *only one* category: "life that is a reproduction."

The only difference, in cases where spontaneous generation was believed, was that the seeds were *invisible!* One could also say that the only difference was that the *source of the reproduced information was unknown.* All life comes from the transmission of information from sources, models, and all cases where spontaneous generation was believed have been refuted when these previously unknown sources were discovered.

Historians of science often present Pasteur's experiments as the last nail in the coffin of spontaneous generation. But this idea is still alive, it only has been pushed back even further. Scientists have proved that spontaneous generation is false both for organisms visible to the naked eye and for microorganisms, but materialists still believe that it was possible for the very first microorganisms, which appeared on Earth billions of years ago.

It is easy to draw parallels between the old beliefs in spontaneous generation and the materialistic beliefs about the origin of life since these are basically the same thing. We can even use the same definition in both cases: it is the belief that organisms can form spontaneously when the right conditions are met.

In materialist theories, the first living systems must be formed by chance. Instead of "by chance," one could also say "spontaneously" or "accidentally," all these expressions are similar. So, it is clear that these theories are only sophisticated versions of the theories of spontaneous generation of yore. In essence, the primary difference between the theories of spontaneous generation of the past and those of today is that, in the past, they did not try to explain in detail how molecules could combine spontaneously to form organisms, while those who believe in it today are trying to do so in their theories.

To try to convince us that their theories are different from the beliefs of long ago, materialists rely on the enormous amount of time that has elapsed since the formation of the Earth. Materialists themselves recognize that the spontaneous formation of organisms is "highly improbable," but since these processes have had millions of years to happen, they believe it gives time for incredibly lucky events to succeed in producing this miracle and start life. In short, the magical ingredient that was lacking in the spontaneous generation theories of yesteryear was a huge amount of time...

This argument is very strange because, if there is one thing that is the great enemy of materialist theories of the emergence of life, it is *time!*

This truth is clearly manifested in the law of probabilities, which we can summarize in these terms: *Everything that depends on chance tends toward its most probable state.* To understand how the law of probabilities work, it suffices to perform a series of coin flips. We all know what the most likely outcome is: around 50% heads and 50% tails since the odds are equal between these two possibilities. Of course, if we make thousands, even millions of flips, it is possible that we get series that are moving away from the average, like a coin that falls ten times in a row on the same side, but that does not change anything in the long term since the law of probabilities tells us that if we continue our coin flips long enough, this extraordinary series will become insignificant and that the average will always remain about 50% for each possibility. This behavior of chance events is well known to all, and it manifests itself through what statisticians call the "regression toward the mean" or the "law of large numbers."

The law of probabilities applies to all domains governed by chance, which includes the different mixtures in which materialists believe that life has appeared. All that a long stretch of time can offer *is an inevitable return to the average* and never the perpetual series of improbable events that the materialists conceive in their imagination! Even if we suppose that it is possible for fantastic series of lucky events to produce drafts of organisms in those mixtures, time can do nothing else but break them down quickly to bring everything back to a more probable state because time never does anything else! A phenomenon that is also linked to the concept of entropy and the fact that energy always tends to spread evenly, those are only different consequences of the law of equilibrium that we have seen earlier.

Time is the great equalizer, it inevitably brings everything back to its place. Once again, materialists are reasoning upside down since what they expect from time *is exactly the opposite of what time can produce.* In reality, the phenomena they hope for can never occur because nature does not work magically, but obeys simple rules that never vary.

9.5 THE TRANSMISSION OF LIFE ON EARTH

In the field of biology, one of the most important laws is the law of reproduction. One day, the skeptics will become exhausted fighting against this great law; they will eventually bow to it, finally accepting that it must apply to *all* life forms, even the first terrestrial life forms.

To escape the dead end of materialism, the first step is to accept that life comes from the invisible side of reality. Life must *first* exist on the invisible side *before* being transmitted to the visible side. Science is perfectly clear on this point: There is never any spontaneous apparition of life, only *transmission* of life! To respect the laws, the concept of the appearance of life on Earth must, therefore, be replaced by the concept of the transmission of life on Earth.

This brings us to another big question: *How* was life transmitted from the invisible side? Indeed, even for those who have no doubt about the existence of invisible life, this process may seem very mysterious.

Without surprise, the answer to this question lies in the universality of the laws. To answer the question of the transmission of life on Earth, we must first understand how information transmission works in general, to recognize the great laws that are at work. We have already seen this subject, and summarized the answer in the law of interactions: *Every interaction is an exchange of energy that carries information.* It must be the same for the interaction between the visible and the invisible, this interaction must be conducted by exchanges of energies that carry information.

Once again, it is the universality of the laws that provides us with the answer since there is only one intermediary for all interactions: *energy.* As we have seen in the previous chapters, energy, or light, is the universal intermediary that conveys all interactions. We must simply bear in mind that the word "light" is used here in a broad sense, that it encompasses all particles of the same nature as light, regardless of whether they are known to current science.

It is *light* that brought life to Earth. This is not a revelation because, intuitively, everyone already knows this! Without light, life on Earth would be impossible. Even the materialists give an important place to energy and light in their theories of the emergence of life. What must be understood is that light not only gave the impulse necessary to the formation of the first organisms but also contained the *information* necessary for their formation, information that already existed on the invisible side of reality.

Earlier in this chapter, we separated life into two broad categories, invisible life and visible life, and presented this separation as a simple figure. This schema is not complete since it lacks an essential element: the link between the two. This link is provided by the flow of energy. Just as the visible and invisible aspects of the human being are connected by two streams of information flowing in opposite directions from each other, the visible and invisible parts of nature are also connected by two complementary currents, which allow information exchange.

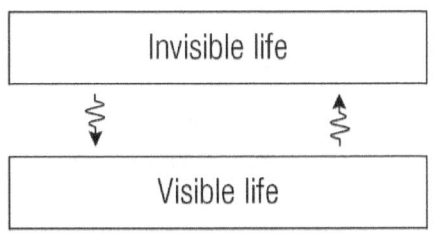

Visible life and invisible life are interconnected by complementary currents that transmit information from one level to another.

The descending current, which goes from the invisible to the visible, is the one that interests us in the case of the transmission of life because it is the one that transmitted on Earth the information necessary for the formation of life. In other words, it brought the *organizing pressure* that countered the disorganizing effect of randomness.

The notion of organizing pressure is very important, and is another notion it applies to a wide variety of areas. For example, when we do the housework, we apply an organizing pressure to our environment, arranging the objects in a specific order that would never occur if the organization of our house was left in the hands of chance. It is the same thing when we renovate our house or repair something; we exert an organizing pressure, which intends to counter the inevitable tendency toward disorder that occurs with the passage of time. It is also the way companies and factories operate, where precise instructions from the management must be respected for the systems to function properly. The most beautiful examples of this principle are found in art, where the organizing pressure, expressed by the artists, can bring out extraordinary works from raw elements, like a statue from a block of stone, a painting from paint, a novel from ink, a symphony from vibrations of the air...

What characterizes an organizing pressure is the fact that it follows instructions, a vision, ideas, models, plans... The impulses it transmits are not random but intended to accomplish a specific purpose.

Ultimately, the effects of an organizing activity are always forms of reproduction. *The processes are always about getting information from one domain, and replicating it in another domain.* The sculptor, the architect or the painter all try to reproduce, as faithfully as possible, what they see in their imagination; the businessman seeks to reproduce in real life the different stages of the business plan he has designed on paper; when we clean up, we try to replicate our idea of what a clean house is; when we cook, we reproduce a recipe, and so on.

Reproductions are not necessarily identical to the model because each area offers different possibilities, but the basic process remains the same. For example, when a studio decides to adapt a novel for cinema, it is a form of reproduction, even if the initial form and the final form are very different. It is the same thing for the information that is displayed on our computer screen; it is a reproduction, in the form of pixels, of the information present in the form of bits on the hard drive of our computer.

Within organisms, it is the genes that play the role of a plan, a program or a recipe. The whole organism is present in the form of codes in the genes, and the organizing flow that passes through the genes is intended to reproduce the organism at another level, in the flesh. In the same way that a computer uses the codes present on its hard drive to form an image from pixels, the codes present in the DNA serve to form the organism from molecules. An organizing current must constantly be present for this process to continue without interruption. For the body, this flow of precise information is just as important as food, water and air; if it is interrupted, death is not far away since there is nothing left to counter the disorganizing effect of randomness.

All the activity of life is an endless assembly line, an endless series of reproductions, connected to each other by exchanges of information. All the difficulties faced by the materialist theories of the emergence of life stem from the fact that, in the beginning, there is *no* organizing pressure according to them, only random activity, which cannot organize anything!

In the universalist vision, the organizing pressure is present from the earliest beginnings of life on Earth since it comes from the invisible side of nature. From the beginning, information-carrying impulses can organize the elements in a determined order, gradually forming organisms as we know them. Moreover, the fact that an organizing pressure is present from the start means that the principle of natural selection can act from the beginning, to participate in the evolution of the organisms, since natural selection comes from the encounter between the organizing pressure of the living and the selective pressure of the environment.

In summary, there were initially invisible models, these models transmitted information through energy, and this information was used for the formation of visible reproductions. This is how the information that existed in the invisible domain was reproduced in the visible domain, and that life appeared on Earth.

Many will think that it is not really a scientifically valid solution to the question of the origin of life on Earth, since we only give an outline and no specific details. However, it is not the amount of detail that makes it possible to judge the value of a theory. For example, we do not accuse physicists, who summarize the functioning of the universe in formulas containing only a few characters, of not being scientific because they stick to the main lines! On the contrary, the more the physicists simplify their formulas, the more they remove the superfluous, and the more the scientific community applauds.

With each detail that is removed, the closer we get to the truth in its purest form. It is the same here, these solutions contain only what is *necessary* to solve the question of the origin of life on Earth, and nothing more! Universalist solutions are simple, but they are not simplistic; they are simple because they focus only on what is essential.

Even if we do not know what exactly happened when life was transmitted on Earth billions of years ago, we know one thing for sure—the laws have not changed! The reproduction processes of the far past followed the same laws as those of today, the only difference being that the sources were invisible, which is only a detail from the point of view of the laws.

People who insist on knowing all the details greatly underestimate the true richness of nature. Nature is inexhaustibly rich and complex; remember that biologists cannot even understand all the processes taking place within a single living cell! They cannot even fully understand how the cells of our own bodies reproduce because it is a process involving thousands of proteins, all kinds of small molecular machines that combine their activity in multiple ways, so much that our poor intellect is quickly overwhelmed when it tries to follow all this movement.

It is the same for the processes involved in the transmission of life on Earth. Even if invisible elements come into play, these are living processes obeying the same rules as other living processes, and if we could study them, we would discover the same inconceivable complexity. We do not gain a clearer vision by going into detail; on the contrary, an accumulation of details can lead to confusion, as this may make us lose sight of the fact that it is always the same laws that manifest themselves through this incredible variety of phenomena.

That being said, there is one point that can be emphasized to get a more concrete picture of this process of information transmission from invisible reality. This point is the essential role that liquid water must have played in this process.

Intuitively, everyone understands that energy and light have had an important role to play in the emergence of life on Earth, and it is the same for water; intuitively, we know that life on Earth and water are linked. Life, as we know it, is inconceivable without liquid water. All theories assume that life was first formed in water, and the presence or not of liquid water is the first question that astronomers ask themselves when they wonder if life is possible on an exoplanet they have just discovered.

Everyone agrees on this point: water and light are two essential elements for life on Earth. It is probably liquid water that has served as a *bridge* between the invisible and the visible because water is both sensitive enough to obey the impulses of light and dense enough to transmit them to the rest of matter. These two qualities of water, its *sensitivity* and its *density*, make it the perfect intermediary between visible and invisible reality. Because of that, liquid water has certainly played an important role in the transmission of life on Earth.

9.6 THE ORIGIN OF LIFE

The existence of invisible life and its transmission in the visible domain, through energy, is a natural solution to the problem of the origin of life on Earth. But this inevitably brings another question: Where does the invisible life come from? Indeed, if we do not answer this question, the enigma of the origin of life is not really resolved since then we have only displaced the problem.

To truly answer the question of the origin of life, it is necessary to draw a portrait of the invisible much larger than what has been done so far in this book and, therefore, to speak of realities that are well beyond what is accessible to the instruments of science. Again, it is the laws of nature that allow us to do so because, even though we do not have direct access to these invisible worlds, we can still have an idea of what they contain by making logical deductions based on the universality of the laws. As always, the key is in the great laws. To solve the great questions of life, one only needs absolute confidence in the universality of the laws and follow *to the end* the logical implications that come with the laws.

One of the main laws of the living is the law of reproduction. This law is universal, so it must apply to all life forms, without exception, which includes the invisible life forms. Therefore, the logical consequence of this law is that the invisible life itself must be the reproduction of another life, and that this previous life must be the reproduction of another life, and so on...

At first glance, this seems to be a problem. If life must *always* be preceded by another life, it seems to mean that the chain of life must regress to infinity, that it has no origin, which does not give us a satisfactory answer. Life must logically have an origin, like all phenomena, but we must first understand that the path that goes back to this origin is *very* long and that there are many levels to go through before arriving at this point of origin.

So far, we have presented the invisible life as only one category. This does not accurately depict this domain since, in reality, the invisible life must exist on many levels, like everything else. Between each of these levels, currents flow and reproduce information from one level to the next.

The existence of life on many levels is only the continuity, in the invisible realms, of the richness of nature that we see all around us. The most natural way to conceive these other levels of existence is to see them as different layers, similar to the structures formed by successive levels that are found in nature, such as geological layers, atmospheric layers, the different tissues of living organisms, and so on. These levels are all formed by the law of selection, which is everywhere at work.

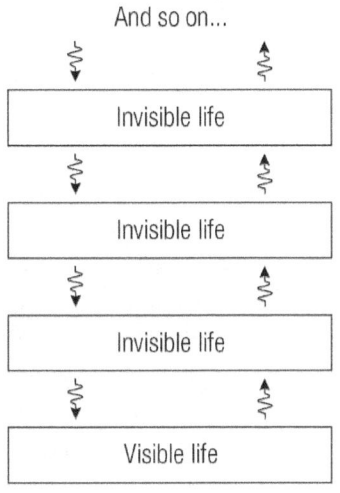

Like everything in nature, invisible life exists on many levels, formed by the law of selection, interconnected by currents that transmit information from one level to another.

The invisible worlds are only successive layers. When we put all these invisible levels together, we get a much grander portrait of reality than what we are able to perceive, but again, that is only the continuity of what we observe in nature, where the tiniest strand of grass contains a complexity that exceeds our understanding. Simply put: *Nature is much richer than we think.* It is only necessary not to lose sight of the fact that it is always the same simple laws that are at work behind this inexhaustible variety of phenomena.

Nature is structured in layers, connected to each other by currents, like a dress is formed by layers of fabric connected together by threads. Regarding the invisible, instead of levels or layers, one can also speak of plans or worlds; the principle remains the same. On each of these levels exists a life that served as a model for the life that succeeded it on the lower levels. These levels can be seen as the bars of a ladder and, going up this ladder, one gradually approaches the primary source of life.

To get a clearer view of what is on the higher levels of this scale, we must then take another step similar to what we had to go through when we were interested in the question of consciousness. In the previous chapter, we saw that we perceive the world only through light, and that means that our center of consciousness must itself be made of a kind of light. This is similar reasoning that must be followed with respect to life. Energy, light, is inseparable from life because it could not exist without energy. It is on this element indispensable to life that we must rely to explain the origin of life, and not on matter, which only plays a secondary role. Like consciousness, *life comes from energy.*

We must put energy at the center of our conception of life, just as we must do for consciousness. This, considering that the light itself can serve as a habitat for life and that it did before the beginning of life in visible and invisible matter.

By seeing life in matter as the reproduction, in another form, of a life that first existed in energy, we obtain a much more coherent vision of the world, where energy always has the first role. Once again, it does not matter if one understands exactly how life can exist in the different kinds of light, what matters most is to place the elements in the right order.

By simplifying the universalist vision of life, we can divide life into two broad categories: life in energy and life in matter. Each of these two huge categories can be divided into several sublevels, each forming a world in itself. Visible life exists in one of the sublevels of the material part of nature, all the rest of life is invisible.

This very broad vision of reality allows us to get even closer to the origin of life, but this portrait is not yet complete. Indeed, some might then assume that this means that life has appeared spontaneously in energy, just as materialists believe that life has appeared spontaneously in matter. Again, such an idea would be false because it contradicts the law of reproduction! All life forms are reproductions, as much life in matter as life in energy, there can never be an exception to this rule.

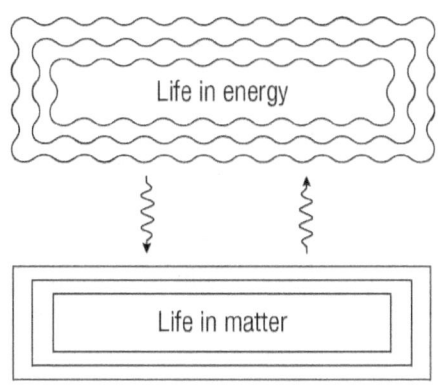

Life can be divided into two broad categories: life in energy, and life in matter. In both cases, life exists on several levels.

To find the origin of life, it is necessary to go back even higher, higher and higher, and continue until the beginning...until we reach the origin of all that exists. This is where the path ends, and that is where the answer to the question of the origin of life lies. The law of reproduction does not mean that the chain of life is without beginning, *but that the origin of life and the origin of existence are one.* This is where the ultimate logical consequences of the law of reproduction lead us.

What lies at the point of origin of all that exists? This is another big question. Reflecting on this question is inevitably talking about God, a subject that is taboo in the field of science. We are constantly told that the question of God is outside of science, that we must turn to religions to reflect on this question, but when religions are examined on this subject, we find many inconsistencies that can only put off those who always seek to remain logical.

The God of whom we are going to speak here has little to do with the vision religions usually present to us, nor is God a bearded man living in the clouds, or any other caricature made to discredit this fundamental notion.

Here is a simple and clear definition of God: *the point of origin of existence.* He is both the point of origin of existence and the point of origin of life. In the context of this book, believing in the existence of God simply means believing that the origin of existence and the origin of life are one and the same. This, not to follow any religious or spiritual philosophy, but, as always, to follow the principle of universality of the laws of nature. The law of reproduction is absolute, it must apply to all life forms, without exception. It means that life is always preceded by another life, and that going up the ladder of life, through the multiple levels of reality, we inevitably come to a point where the origin of life merges with the origin of existence, which means that they are *one!*

It is a question of logic: the fact that the law of reproduction is absolute means that the origin of existence and the origin of life are necessarily the same thing. The answers to all the big questions are in the universality of the laws of nature. The same is true of the proofs of the existence of God, they are in the pure logic of the natural laws.

It may seem futile to approach the question of God since we cannot really understand what is at the point of origin of existence. But one must not be discouraged by the fact that one cannot understand everything about it, nature is overflowing with realities that the intellect is unable to grasp perfectly since it remains an instrument with very limited capacities. Even with the best efforts, our intellect can understand the world only partially, forming representations that are always limited. This rule applies all the more when thinking about God. He cannot be fully understood, and trying to describe him with words is like trying to cover the Sun by throwing handfuls of ash at it!

This being said, it is still possible to understand certain aspects of what must be at the origin of existence, even if this understanding must be limited to symbolic representations. For example, it is possible to represent the Sun using a circle since certain aspects of this star are correctly represented by the image of the circle. Therefore, the circle is an adequate symbol to represent the Sun, even though we know very well that the difference between reality and our little drawing is unimaginable! It is the same regarding God, it is possible to represent him imperfectly with the help of symbols.

Thanks to the logic of the laws of nature, it is easy to deduce what must be at the point of origin of existence. Physicists have already known this for a long time: what is at the origin of existence is a *force*. This is basic logic because, without force, no phenomenon is possible; therefore, the existence of a force must necessarily precede all that exists!

There must be the existence of a force before anything happens, including the birth of the universe. Physicists who study the forces of nature recognize this fact because, according to their theories, what existed at the origin of the universe was a *unique* force, which quickly divided, after the birth of the universe, into the currently known forces.

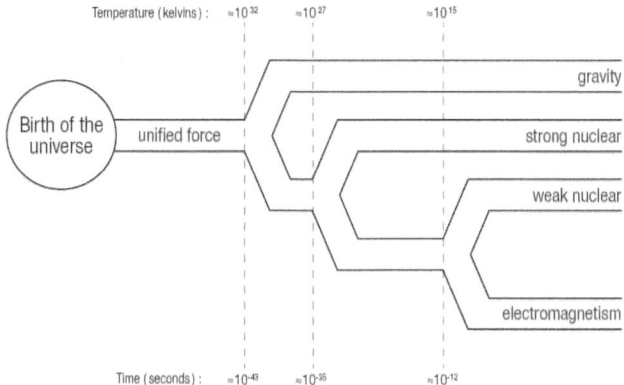

This does not mean that this force that existed at the birth of our universe, the so-called Big Bang, is truly the original force. To avoid conceptions that are too limited, one should rather see the Big Bang as a stage in a chain of events that go back even further in time. Be that as it may, the principle must remain the same; the closer we come to the origin of all that exists, the more the apparently divided forces become one. This, up to the point of origin of existence, is where all the forces are united in a single force: *God*.

In addition to the "point of origin of existence," God can, therefore, be defined as the "unique force," the original cause of all movement, and we can see the whole of existence as an immense chain reaction caused by this unique force.

This reflection brings us to a conception of life that is completely different from the one that materialists propose. Indeed, if the origin of existence and the origin of life are the same thing, and if what is at the origin of existence is a force, it logically means that *life is a force*. This means that what is commonly called "life" is not really life, it is only a succession of *effects* due to the existence of life, which is the unique force behind everything that exists!

Some will say that this reflection goes too far, that it is only philosophy and not science-based thinking. This is false since the guiding thread of this reflection is always the implacable logic of the laws of nature, laws that have been validated by innumerable scientific experiments. We only follow *to the end* the logical consequences that result from the laws. This gives us a vision of reality completely different from the one proposed by the materialists, but a vision perfectly consistent with science.

To go further, we must now add another essential element to this portrait: information. When we dealt with the question of the origin of life on Earth, we saw that everyone already considered light an essential element of life, an element that provided the impulsion, and that it was also necessary to admit that light had also provided the information for the formation of the first terrestrial organisms, to obtain an important key. It is the same thing with the unique force, Life with a big "L." Everyone can understand that what is at the origin of existence is necessarily a force, that it is this force that gives the impulse to all that exists. To obtain a complete picture, one must also accept that this force has given not only the first impulse but also the *first information,* and that it is this original information that is *reproduced* to infinity through the multiple levels of reality. Therefore, at the origin of existence is not only a primordial force but also primordial information—the two are inseparable from each other.

These are broad and abstract concepts that may seem hard to understand at first sight... It is true that we touch the highest realities on which it is possible to think, but that does not mean that these concepts are difficult to grasp. To solve the big question, we must *simplify our thinking* as much as possible, because only in this way can one see reality in its purest form. It is at this level that we find the great answers, answers the simplicity of which is matched only by their beauty.

The higher the concepts, the simpler they are. It is always like this, the higher you get, the simpler things become. For example, when we are on the surface of the Earth, it seems to be the most complex object. But, if one rises with the aid of a plane, everything becomes increasingly uniform, and if one continues to rise more, the Earth itself becomes a simple sphere, and then a single luminous point. All the incredible variety that we observed before is then contained in a point, the simplest object that is!

This is how one can conceive of the primordial force and the primordial information, as a single point, perfectly immutable, from which emanates the whole of reality. The point is a symbol that can be used to represent God, just as the Sun can be represented by the circle because the point is the purest figure that can be conceived. Moreover, this idea is already contained in the definition of God given previously: *point* of origin of existence.

Just as the seed of a fruit is a reservoir of energy and information, potentially capable of producing a magnificent tree, God can be seen as a point where the original force and information are gathered, which serve to form all worlds. Everything functions according to the same immutable laws, and what we observe, every time a seed germinates before our eyes, it is only the repetition of what happened at the origin of existence, and which is constantly reproduced under an infinite variety of forms.

This primordial force and information are Life itself, the perfect and eternal model of which all things are imperfect reproductions. Instead of information, one can also speak of instruction, rule or law. The pure coherence that we find in God is the primordial information, and this perfect equilibrium is also the fundamental law of nature, which indicates to all things how it should behave.

Some might believe that there is a contradiction in what is said here because we constantly repeat that all life forms are reproductions, but that we do not apply this law to God. One could then believe that we make an exception to the law of reproduction, which is not the case. The law of reproduction applies to all *forms* of life; consequently, in the domain of life forms, there is never an exception to this law. But God does not belong to this domain since God is not a life form, but Life itself!

Instead of life forms, one could also speak of manifestations of life or effects of life, these are similar expressions. God is not a form, a manifestation or an effect, but the *cause* that precedes everything that exists. This does not mean that we must consider that God is outside the natural laws, but rather that God is the *point of origin* of the laws. He is the eternal model upon which reality is built, as well as the force that engenders and maintains the movement that animates the whole of existence.

Everything is constantly trying to reproduce the harmony, the balance, the coherence that we find in God. Every movement, from the vibrations of the atoms to the dance of the planets and the galaxies, is never anything but a displacement toward a point of equilibrium, that is to say, a point where the effects of the different forces of nature are equal.

The fundamental law of nature is the *law of equilibrium.* The law of reproduction, as well as all the other laws, are only consequences of this law, like different branches connected to a single trunk. At the moral and spiritual level, this law can also be called the law of Love, or the law of Justice, since it is this law that demands a balance between giving and receiving. It is simply the law of life, the Perfect Law, which must serve to understand the functioning of nature, as much as to guide our lives.

All phenomena tend toward a state of equilibrium that is as perfect as possible. Atoms, molecules, crystals, organisms, ecosystems, cultures, civilizations, planets, stars, solar systems, galaxies...all structures are born of a state of equilibrium between different forces, and the currents circulating between the elements are only intended to maintain this equilibrium. Intuitively, we all recognize the importance of this law since we see beauty in symmetry and balance. We are spontaneously attracted to this beauty because we feel that this is something ideal toward which we must tend.

The great difference between the equilibrium that exists in God and the one that is accessible to us is that the equilibrium that exists at the point of origin of existence is a *perfect* equilibrium, whereas the equilibrium found in nature is always a *dynamic* equilibrium between a wide variety of elements. Nothing is ever really in equilibrium in nature, everything oscillates around a point of equilibrium that is never truly reached, and it is this impossible pursuit of the perfect balance that keeps the universe in a state of perpetual motion. In nature, balance must be *constantly maintained*, and it is endless work.

It is impossible to describe all that is covered by the law of equilibrium, which encompasses the whole of existence. It is this law that is the answer to the question of the origin of life. What is at the origin of life is not a succession of random events, as imagined by materialists, but a *LAW*, the law of equilibrium, which can also be called the law of life. A law that is ultimately Life itself, the unique force, God, the point of origin of existence.

Every organism is born, lives and dies only by this law because each creature can only survive if it succeeds in maintaining a certain state of equilibrium. An organism is a state of dynamic equilibrium between different elements and nothing else. Whether it is body temperature, blood pressure, hydration, degree of acidity or something else, the maintenance of the body depends on its ability to maintain a vital balance between a wide variety of elements by a process that biologists call "homeostasis." When all the elements oscillate around equilibrium, the organism is healthy, and the more they move away from it, the more the organism is close to death. The same rule that applies to one organism also applies to ecosystems and societies, which also depend on maintaining a delicate balance between a wide variety of elements.

Within an organism, every impulse, every cell, every organ exists only for the purpose of maintaining the vital balance. To obey this law, each part must be at the service of the whole, and the whole must be at the service of the parts. It is because they obey this perfect rule of functioning that we can be summarized in the phrase "one for all, all for one," that organisms are extremely coherent structures.

The law of equilibrium, which governs the functioning of organisms, is at work behind every phenomenon. It is the responsibility of everyone to seek to understand this law better, through the experiences of their own life. This is the most important work since we cannot neglect the law of equilibrium without this having serious consequences. Our individual health, as well as the survival of our civilization, literally depend on our ability to understand and comply with this law! All the social, economic and ecological crises that we are currently witnessing stem from the breakdown of various equilibria, the importance of which we have neglected. And the consequences of the chain reactions that we have initiated, in our incredible collective stupidity, will still be coming for a long time....

One could go on indefinitely about the law of equilibrium, the law of life, since it is an inexhaustible subject—everything is included in this law! We will conclude this reflection on the origin of life by looking at other symbols that will allow us to summarize what has been said so far.

We can conceive of God as the central point of existence, the point of origin of everything. A symbol that can be used to represent this point of origin is the cross with equal branches, one of the simplest symbols, but also one of the richest.

The cross is a very ancient symbol, its use goes back to prehistory, well before its appropriation by Christianity. In addition to its use to represent the sacred, it is also the best way to represent a central point or a point of origin. Moreover, mathematicians and physicists constantly use this symbol in this sense. When they want to visually represent the result of their formulas, first they draw a cross, that is to say, axes that intersect perpendicularly, on which they place coordinates. The place where the axes intersect is called, rightly, *the origin.*

In this book, the cross represents the point of origin of existence, as well as the harmony that emanates from this center, the perfect balance

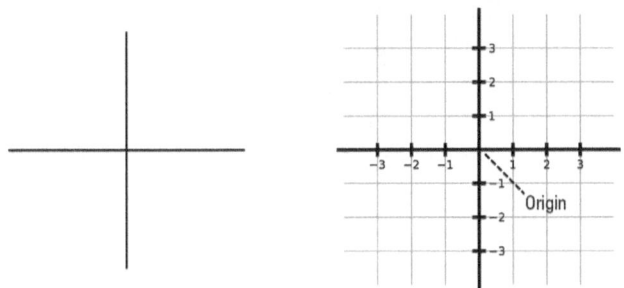

The cross with equal branches is a very deep symbol. It is also one of the most used symbols by mathematicians and physicists.

between the complementary principles, male and female, represented by the vertical line and the horizontal line.

The two arms of the cross also represent the complementary currents of action and retroaction that traverse the whole of existence, such as the arterial and venous currents that sustain the organisms, the currents of will and consciousness that connect the spirit and body, and the descending and ascending currents that connect the different levels of reality. We can also see the arms of the cross as the laws of nature, which are the action of the unique force situated in the center, an action that always acts in a straight line and uniformly from the origin.

Thus, the cross with equal branches represents God, as well as the perfect order imposed by God, through the laws of nature. This symbol represents the foundation of existence, the guidelines around which reality is built. Therefore, it is a perfect symbol to represent the origin of existence as well as the origin of life. This completes the figure we have been gradually constructing since the beginning of this chapter (top figure).

Between us and the origin of life there is an inconceivable distance, this space is not empty but filled with a life that exists on a great variety of levels, each more wonderful than the next. On each of these levels, an incredible diversity of species combines their activities and form ecosystems, the complexity of which is beyond comprehension, as is the case here on Earth. At first sight, this variety may seem

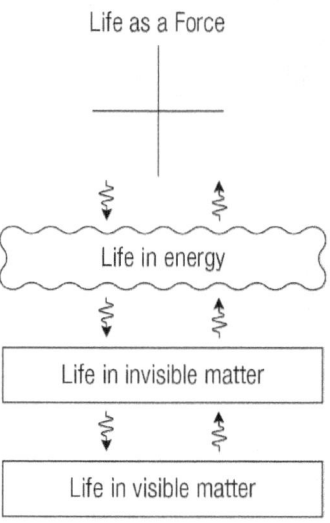

Beyond life in visible matter, there is life in invisible matter. Beyond life in invisible matter, there is life in light. Beyond life in light, there is Life itself.

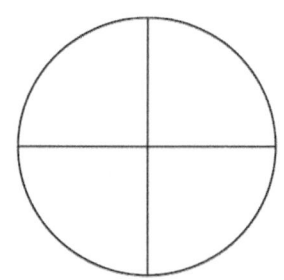

The center represents God, the unique force, Life. The arms of the cross represent the action of the unique force, the laws of nature, as well as the complementary currents. The circle represents the combination all the worlds formed by the laws.

incomprehensible, but it is always the reproduction of the same perfectly simple pattern.

To visualize what this motif is, this information repeated infinitely, we can place the cross in a circle (bottom figure of the previous page).

Once again, the center represents God, the unique force, Life. The arms of the cross represent the action of the unique force, the laws of nature, which generate the complementary currents that maintain the cohesion of the whole. As for the circle, it represents the whole formed by the laws: reality, nature, the visible and invisible worlds.

The depth of this symbol is inexhaustible, *everything is there!* From the beginning of this book, we have emphasized the fact that the role of the intellect is to form representations of reality, which serve us as maps to understand nature. The figure of the cross in the circle is one of the highest representations that can be conceived since it is a figure representing *the whole* of reality.

This is a figure reduced to the essentials, but this figure contains *everything*. In the same way that moving away in space, we can manage to contain a planet, a star or a galaxy within a single point, this diagram represents reality as it would appear to us if we could see it with one look.

This figure, representing the whole of reality, is also the basic model of everything since the parts are only reproductions of the whole. All that exists is only a reproduction, with variations, of this basic structure. An atom and its nucleus, a cell and its genes, a fruit and its seed, a solar system and its star, a galaxy and its center, an organism and its heart, a body and its spirit, a wave and its impulse—*everything is there.*

The law of reproduction is also represented in this figure: the point is the model, the circle is the imperfect reproduction of the point, and the arms of the cross are the streams of information by which the reproduction is made. Again, one could go on indefinitely on the symbol of the cross in the circle since it summarizes the structure and functioning of nature as a whole. Everyone is invited to dive into this symbol to discover its infinite depth.

As final examples of the richness of this symbol, here are other figures that give us different ways of representing reality as a whole (next page).

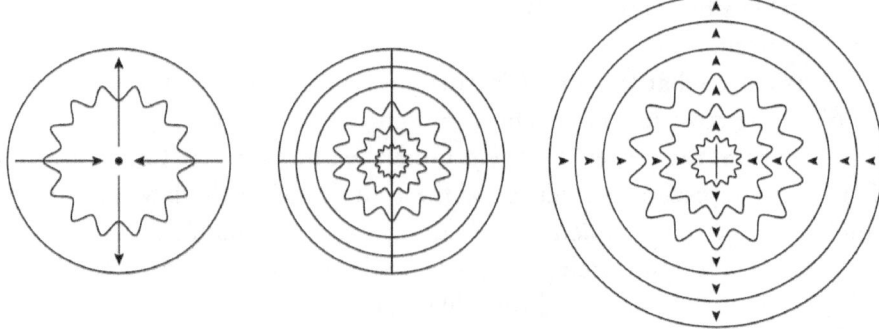

Once again, there is the point of origin of existence at the center; the levels represented by the undulatory lines are the levels constituted by the different kinds of visible and invisible energies; the levels represented by the simple lines are the levels consisting of the different kinds of visible and invisible matter; and the two arms of the cross represent the complementary currents that traverse the whole and maintain its cohesion, currents that take different forms according to the levels where they manifest themselves.

The number of levels here is arbitrary, the goal is only to represent the basic notions. On each of the levels of nature, there exists a life with different possibilities, a life that is increasingly more intense as one approaches the origin, the point where resides Life itself. Those same figures can also be used to represent the structure of the human being, as we saw in the previous chapter, since it is also only a reproduction of the whole.

On its own, the figure of the cross in the circle summarizes the whole of universalist philosophy. As everyone can see, the principles that support this theory are perfectly consistent. Everything is summarized by these simple and elegant lines. The figure of the cross in the circle contains the main answers to the big questions of life, *everything is there.*

9.7 IN SUMMARY

The keys to solving the mystery of the origin of life are in the universality of the laws, the invisible and energy, just as the keys to solving the mystery of consciousness.

Universalism allows us to progress toward an ideal of logical coherence since it allows us to solve the big questions by relying on a *single* principle:

the universality of the laws of nature. Everything follows logically from it: the existence of the spirit, of invisible worlds, of invisible life... The existence of all this becomes a logical necessity as soon as we accept the logical consequences imposed by the universality of the laws; in other words, as soon as we understand that the laws allow absolutely no exception.

What we need to seek is an increasingly higher level of conceptual unification, which allows us to explain more phenomena using fewer laws. This unification allows us to reach ever-higher levels of understanding, to see more clearly. Just like the universalist explanation of consciousness, the universalist explanation of life frees us of the artificial divisions put in place by our intellect. This confusion is replaced by a vision where everything obeys the same laws, a more coherent and elegant theory.

Universalism is the simplest philosophy there is, so much that it is not even necessary to pronounce a single word to present it: it is entirely contained in the image of the cross and the circle. This figure summarizes the structure and functioning of nature as a whole. It is enough to reflect deeply on this single image, to understand the essence of what there is to know about life, the universe, consciousness...everything is there!

Here are the most important conceptual unifications present in the universalist explanation of life, which helps us to build this consistent worldview:

The first life forms were reproductions, like all life forms.

The visible life comes from a transmission from the invisible life.

There are life forms in the invisible domains, as there are in the visible domains.

There are life forms in energy, as there are in matter.

Invisible life follows the same laws as visible life.

The origin of life and the origin of existence are one.

Life is the unique force at the origin of existence.

To answer the big questions, we must put the universality of the laws, the invisible and energy, at the *center* of our vision of the life, rather than the periphery as is the case in the materialistic conception. This universalist vision is completely reversed in relation to materialism; it is a change comparable to the transition from a worldview where the Earth is at the center, to another where the Sun is at the center. In the universalist vision, life is no longer a strange exception, but rather a universal manifestation of the laws of nature. All that exists is only different forms of life, different reproductions of Life. The same wonderful richness of life exists both in the visible and the invisible, both in energy and matter, from the origin of all things to the last reaches of the universe.

10. THE UNIVERSAL ORGANISM

In short: everything is life.

A definition of the universality of the laws that we have seen is this: The laws act the same way through all of reality. But it is possible to define the universality of the laws in many other ways, and the definition we are going to explore, in this chapter, is as follows: *Every part is a reproduction of the whole.*

This definition tells us that, no matter what sample of nature we observe, it will always have the same characteristics as the whole of reality. It does not matter where we are in space or in time, whether we are at the scale of atoms or galaxies, in the visible or the invisible, reality always has the same basic structure, as well as the same functioning. Wherever we look, we find only *variations on the same theme,* a theme that is imposed by the laws of nature.

In mathematics and physics, a structure of this kind is considered *self-similar,* a word used to designate an object the parts of which are similar to the whole. Fractals are geometric shapes that have this feature because regardless of the scale at which they are observed, they reproduce the same patterns in a similar way. Holograms also have this quality, since each part contains information about the whole; this is the reason a hologram can be cut into pieces, and each of them will present the same image as that of the original hologram, only with a more limited field of view.

At first glance, the concept of self-similarity may seem to us as something special existing only in certain cases. But when we think about it more thoroughly, we realize that this concept is useful in many areas and that it can help us understand many important things about nature.

Although it is an unusual word, the concept of self-similarity is already intuitively known to all. The most common example of self-similarity is the structure of a tree since each of its branches repeats the same basic pattern. In reality, all living organisms can be considered self-similar structures, since they are formed from the same repeating patterns at all scales. Our body is an organism made up of small organisms, our cells, which have the same basic characteristics as the organism as a whole. Like the example of the tree, it is not a perfect self-similarity, since the cells are not exact reproductions of the body as a whole. But it is still a form of self-similarity, since the principles that direct the structure and functioning of cells are the same as those that direct the organism as a whole, so they can be considered variations of the same thing.

Ultimately, nature itself is a vast self-similar structure, where the same patterns are reproduced infinitely. *Everywhere* we only find the same basic structure that repeats itself, only with some variations.

At the end of the previous chapter, we have already begun to explore this notion, seeing that reality can be symbolically represented by the image of the cross in the circle. This figure represents an overview of nature, and all forms are more or less exact reproductions of this model. All forms tend toward the circle or the sphere, that is, toward symmetry, equilibrium and uniformity; this is because the laws of nature constantly push all things toward a state where the forces are in equilibrium, a balance of the forces represented by the equal branches of the cross.

Reality is born of a state of perfect

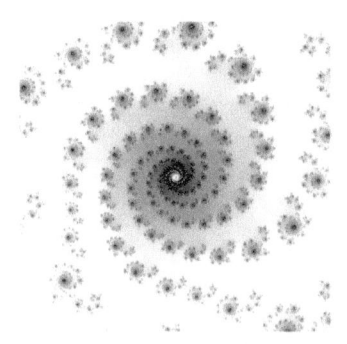

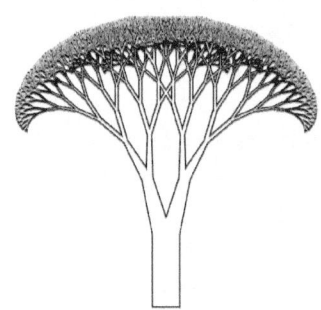

Different examples of self-similar structures. From top to bottom: a fractal, a hologram, a tree.

coherence, perfect equilibrium, a perfection that the laws of nature constantly seek to reproduce. The original perfection is symbolized by the point at the center of the cross, and the circle is an imperfect reproduction of this point. This circle represents reality, nature, while the cross represents the laws, or the forces of nature, that come from the origin and maintain the cohesion of the whole.

Nature is an approximation of the original perfection. Everything is an imperfect reproduction of the ideal model that is at the origin of existence, namely, God. This model is eternal, it has always existed and will always exist, so it is not a reproduction of anything. He does not possess a form that we are able to understand, since all the forms we know are finite, whereas God is infinite. What we can do, however, is to represent God with symbols, and in this case, the chosen symbol is the point.

All the elements that populate nature are imperfect reproductions of this point, this original model. Therefore, all elements try to be as symmetrical as possible, all things are approximations of circles or spheres. Of course, this does not mean that wherever we look, we will only see circles or spheres, rather, that it is a general *tendency* that directs all phenomena.

This is manifest even in the most mundane things. Take, for example, a box of rice: at first glance, we can see a violation of this principle, since this form is not spherical. But this box actually contains a pressure that pushes it to become a sphere, and if this tendency cannot be fully manifested, it is only because there is a resistance that prevents it, constraints that come from the rigidity of the box. If we transfer the contents of the box into a bag, we will immediately see this natural tendency toward the sphere manifesting itself more clearly because the bag will oppose less resistance. Indeed, the rice bag will adopt a shape closer to the sphere than the previous box. If we continue our observations, emptying the contents of the bag on a table, our grains of rice will form a pretty pile with a circular base; once again, the tendency toward symmetry manifests itself, even if, because of the coarse nature of the grains, it is still far from perfect.

Again, the example of the grains of rice shows us the law of equilibrium, a law that causes everything to always be distributed as uniformly as possible. The innumerable particles that make up the universe are like the grains of rice in this example, they too adopt the most balanced configurations possible, a tendency that comes from the fact that the energy that animates everything continuously seeks to distribute itself evenly.

The tendency toward perfect uniformity, which comes from the ocean of energy in which the universe bathes, meets resistance everywhere because there exists between particles different kinds of attraction and repulsion. These interactions push the particles to form atoms, molecules, crystals and all kinds of more or less rigid materials, which, like the box in the previous example, can resist the natural tendency toward uniformity.

Whether it is a box, a bag, a molecule or a crystal, in each of these cases, the configuration is always one that is as close *as possible* to equilibrium. Even if these forms are not perfect reproductions of the ideal state, symbolized by the point, the circle or the sphere, it is always the best *approximation* possible; this approximation is an imperfect reproduction, but a reproduction all the same. So, if the tendency toward uniformity cannot be fully manifested in most cases, it is because there are resistances that prevent it, and not because the law of equilibrium is suspended. It is a blessing that things are like that since if the world was perfectly uniform, it would be very boring!

Another example that seems to contradict this trend is that of a tree. Indeed, at first glance, we can see in the entanglement of branches a structure that does not care about the law of equilibrium. But, if we go further, we realize that on the contrary, the structure of a tree is entirely guided by this law. First, the tree must spread its branches evenly around its axis, otherwise, with time, it will lean to one side and eventually fall. Second, it must also extend its roots underground, in a way that balances its deployment above the ground. Then, if we observe sections of the trunk or the branches, we see that they form circles…

A box, a crystal and a tree are rigid structures, for this reason, the tendency toward equilibrium that is everywhere present cannot be fully manifested. This tendency manifests itself entirely only in environments that offer little resistance, as is the case, for example, with liquid water. Water is an excellent medium for observing the law of equilibrium at work: raindrops spontaneously adopt a spherical shape; when they reach the surface of a lake, the shock wave propagates in a circle; bubbles are also spheres; and so on.

We can also look to the sky to see this law manifest itself clearly. Planets and stars are all spheres, these celestial bodies follow elliptical orbits, and even the galaxies are in the form of disks or spheres.

Everything is a reproduction of the cross in the circle. It means that all the forms tend to recreate the state of perfection that is represented by this symbol, a tendency that is imposed by the laws of nature. The forces of nature seek equilibrium, and as long as this state is not attained, there is a pressure that pushes toward this ideal. Perfect balance being impossible, the forms adopt an approximation that is as close as possible, with a result that depends on the constraints of the environment. Perfection is unattainable, but it is still possible to *approach* it, thanks to the harmony that is created when the forces of nature are in a certain equilibrium, as is the case in a healthy organism, ecosystem or society. It is this state of harmony, of symbiosis, of communion, which the laws continuously seek to reproduce because that is what is closest to the primordial coherence. When an element is in harmony with itself and with its environment, it is perfect in its own way, and this element can then be described as a reproduction of original perfection.

Therefore, we must consider the symbol of the cross in the circle as we consider the equations of physics, that is to say, as an abstract representation that expresses certain fundamental relations of nature, a geometrical figure that shows us what are the great laws that direct the functioning of the universe. This appears to us even more clearly when drawing the cross with arrows or vectors.

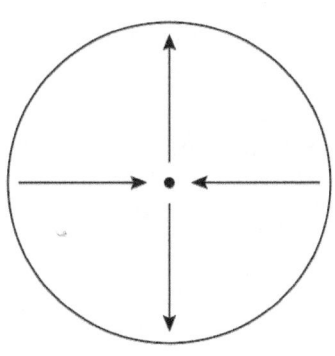

Here, the branches of the cross are shown to us as complementary currents: a current that goes from the center to the periphery, and another that goes from the periphery to the center. This diagram represents many of the great laws we have seen: the law of retroaction in the opposite currents; the law of equilibrium in the balance of the currents and the symmetry of the circle; the law of interactions in the currents of energy that travel between the center and the periphery; and the law of reproduction, since the circle is an imperfect reproduction of the point.

The symbol of the cross in the circle is inexhaustible—*it summarizes all of nature!* To have access to the knowledge that this figure contains, we must first accept that the foundations of reality are clear and simple, and not mysterious and complex. This symbol is the basic pattern of nature, a pattern that repeats itself everywhere, with variations depending on circumstances.

The diagram of the cross in the circle can be used to represent a wide variety of systems: the heart and the body, connected by the arterial and venous networks; the brain and the body, connected by the sensory and motor nerves; a star and its planetary system, linked together by the radiation emanating from the star and its gravitational attraction; the spirit and body, connected by the currents of consciousness and will.

It can also be an atom and its nucleus, a fruit and its seed, a house and its inhabitants, an individual and the society, an organism and its environment, and so on. All elements can be represented using this diagram since each element has a center and a periphery, as well as exchanges between the two, which are done through complementary currents.

From the point of view of the laws, everything is a reproduction of the cross in the circle, even if the external form can move far away from ideal symmetry. The forms are widely varied, but the *main lines* are always the same. This is because the forms are relative, but the laws are absolute. In other words: *The forms change, the laws remain the same.* And the most important of these laws are summarized in the figure of the cross in the circle.

The laws act the same way through all of reality; all parts are reproductions of the whole; nature is self-similar; the forms change, the laws remain the same... These expressions are different ways of defining the universality of the laws of nature, a principle that we must put at the center of our worldview if we aspire to understand the world logically.

Thus, the cross in the circle summarizes the essence of nature: its laws, its functioning and its structure. This figure shows us the main lines of reality, and we can use this diagram to obtain a good general understanding of nature. What the study of this figure allows us to obtain is a perfectly *consistent* vision of reality, and it is this overview that is sorely lacking today.

Reality as a whole is represented by the figure of the cross in the circle, and each part of reality is a variant of this basic motif. Nature is self-similar, we could even say that it is *holosimilar,* meaning that it is similar in every respect. Nature is like a homogeneous mixture: No matter which sample we observe, it is always representative of what we find everywhere else. Even if there are differences in the outer forms, the basics always remain the same.

This way of thinking allows us to answer the biggest questions, since it means we have everything at hand. We can observe any sample of nature to find laws that apply to all the rest, and these laws are all we need to study to find answers to the most important questions, as we have tried to do this throughout this book.

The laws of nature being universal, it means that all the samples are representative of the whole. Therefore, it is *impossible* to find a sample of nature that is not like all the rest. If we follow this reasoning to the end, we get a vision of the world completely different from that proposed by the materialists because it means that the Earth, and all the elements it contains, are representative of what exists everywhere else. Since what we find on Earth is life, it means that there must be life everywhere...

In short: *everything is life.*

This is an assertion that can startle many, but it is just another logical consequence of the universality of the laws. Life on Earth is not a strange anomaly, *it is a representative sample of the whole!* It is a conclusion to which one necessarily arrives as soon as one accepts that the universality of the laws is *absolute*. There is *nothing* that is not representative of the whole, and it is the same for life on Earth. It simply tells us that the whole of reality is life! This is a conclusion that is not a revelation, since it is an idea that has accompanied humanity since the dawn of time.

Life on Earth is a tiny sample of the life that permeates *the whole of existence*. Life, nature, reality...it is the same thing! All that can be observed is part of life, and so everything is just a different form of life. It is necessary to abandon this artificial division between the living and the non-living, so dear to the materialists, to see everything only as different *levels* of life, forming parts of a vast scale of which humanity knows only a few degrees.

Again, this way of seeing allows us to solve certain artificial mysteries since it makes the enigma of the distinction between the living and the non-living disappear. Indeed, even if one of the most important branches of science, biology, has developed around the subject of life, there is still no consensual definition of life! There are dozens of different definitions of life, and whenever a specialist thinks he or she has found the boundary between the living and the non-living, another expert finds cases where this definition does not work. This is a problem similar to the enigma of the brain, where there is no consensus on the threshold that would allow this organ to generate consciousness; in both cases, it is because one believes in the existence of a limit that is purely imaginary!

When we understand that the law of reproduction applies to all that exists, that everything is a reproduction of the same basic pattern, this idea that everything is life then appears to us as quite natural. Atoms, molecules, crystals, minerals, plants, animals, humans, rivers, oceans, atmosphere, planets, stars, galaxies, the universe...we find everywhere reproductions with variations of the same basic pattern, symbolized by the cross in the circle. The more we deepen this universalistic vision of the world, the more we realize that separating these forms into "living" and "non-living" is a futile exercise. Everything is part of the same living tissue, of the same universal organism, within which everything is interdependent, as are the different organs of our body.

Instead of a universal organism, one can also speak of universal organization, of universal order, or even of Creation, as religions do. All these expressions are similar. The same order is present everywhere, allowing variations in the details, but not in the principles; and the best way to understand this universal organization is to study an organism like the human body.

For example, the heart and the brain are different organs at a certain level, but on another level, they are inseparable, since the heart cannot exist without the brain, and vice versa. Indeed, if the brain, seeing this organ so strange and different that is the heart, decided to get rid of it, it would quickly have a surprise since it would result in causing its own death! Within an organism, everything is interdependent, so much that if one part of the organism has difficulties, it has negative consequences on all the others. This is because each organ is part of the same living tissue, and the fruit of each one's work is necessary for the well-being of all others.

This interdependence extends beyond organisms since none of them is self-sufficient. Each creature depends on the ecosystem of which it is a part, and because of this, an ecosystem can be considered a living tissue. Again, this tissue has a different shape than the tissues of the human body, but the laws that animate it are identical, they are only variants of the same thing. Within an organism, it is impossible to harm some part without this having a negative impact on everything else; likewise, it is impossible to act on one part of an ecosystem without affecting the other parts. This is why, when a species disappears, it often results in the disappearance of other species because they rely on each other.

The interdependence extends to the whole universe: Everything owes its existence to something else. *Nothing is self-sufficient.* Planets owe their existence to stars, and stars owe their existence to the generations of stars that preceded them, going back to the birth of the universe. Even the carbon of which our body is made and the oxygen we breathe are elements that were formed in very old stars that existed long before the Sun. The stars are like organs, within which elaborate processes occur, which give birth to almost all the elements we know, and these stars go through processes of birth, maturation, death, seeding and reproduction, like everything else. Stellar systems can be considered living cells, within which the elements are transformed, before being distributed in space, through processes that span billions of years.

Of course, the living processes that take place on the cosmic scale have a different *form* than those that take place on the scale of a cell of the human body, but the *laws* remain the same. Nature is a universal organism, obeying the law of interdependence, which can be summarized as follows: *Each part serves the whole, and the whole serves each part.* Inside the universal organism, everything is alive because everything is animated by this unique force that is Life. This force is at the origin of all that exists; it is the only thing that is truly independent, and it acts as a perfect law from which all other laws originate.

There is only one thing, LIFE, and it has *two poles.* That is, life as a force, which can be called Life with a capital letter, or God; and life as phenomena, as forms, that is, nature or reality. This is a vision summed up in the image of the cross in the circle, where the center represents one pole, and the circle the other pole.

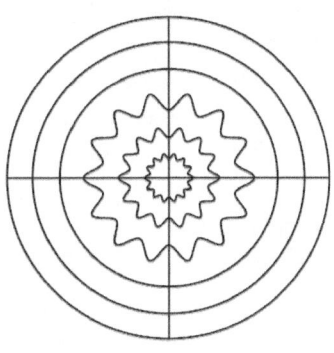

If we have difficulty conceiving of nature as a universal organism, it is also because we do not perceive most of the life that it contains. Life exists on many levels, and the majority of these are invisible to us. Trying to understand the cosmic organism by studying only its visible part is like studying a creature by observing only its skin, its outer layer, its periphery. To have a correct vision of the life that permeates the universe, we must add many other levels to those that current science knows, and it is only when we combine all these layers that the portrait becomes just.

Reality is made of different energetic and material worlds, visible and invisible. All levels are traversed by the same fundamental laws, symbolized by the cross with equal branches, laws that are the manifestation of the unique force, Life, located at the point of origin.

The number of these layers is of little importance here, since we focus only on the general principles. What is important is to understand that the life that inhabits the invisible planes is not supernatural; it is only variants similar to the life we know since the laws allow absolutely nothing else. This pattern also applies to the human being who, in addition to the spirit, possesses many other invisible levels, which can be called envelopes, bodies, vehicles or instruments.

Reality is composed of superimposed layers, and each of these degrees is like a living tissue, composed of interdependent elements. These layers can vary greatly in their composition, but the principles remain the same. By saying that the invisible contains several levels, we are only saying that the invisible is like the visible, that it is structured in a similar way.

Science is wary of the appearances, and universalism too. What is essential is to always respect the great laws, the same laws that are the basis of the primary branches of science, such as physics, chemistry and biology. The universalist approach places the laws above the appearances, and, of course, it gives us a vision of the world that is inverted in comparison to materialism, since in this philosophy, it is the appearances that are above the laws!

In the materialistic vision, the universe is essentially empty and sterile, while in the universalistic vision, it is full of life, because everything is part of the same universal organism. To see this big picture, we must consider that each level of reality is a reproduction of the previous level, going back to the origin of existence, to God. The levels or worlds that formed first are those that most faithfully reproduce original perfection; they are the worlds of light, the highest planes. These worlds naturally adopt forms close to perfection because the substance of which they are made, of a nature similar to light, does not resist the pressure of the original force, which pushes all things to vibrate in harmony. It is in these worlds that live the entities that serve as primordial models for the innumerable species that later formed on the other planes, entities whose humanity already has some knowledge, calling them "gods," "spirits" or other names. Again, there is nothing new in what is said here, since the idea that, in the higher planes, entities exist that are the personification of particular qualities or virtues, and that serve as models or archetypes, is a notion found in many belief systems, especially in polytheistic religions. This is another idea that has accompanied humanity since the dawn of time.

The further we move away from the origin, the more the worlds are gross reproductions of the original perfection because these lower worlds are copies of copies of copies... Contrary to the worlds of light, the material planes offer resistance to the pressure emanating from the original source, so that the processes that occur rapidly on the higher planes must follow a slow evolution here below. Despite these differences, the reproductive processes remain fundamentally the same across all levels of existence...

Once again, this is a portrait brushed in broad strokes. But those who will take the time to deepen the ideas sketched out here will see for themselves that they can obtain a complete vision of reality, a vision that fits as much with the great laws discovered by science as with the great revelations that have been offered to humanity, since both are only different ways of speaking about the most important realities.

11. SCIENCE AND REVELATION

The two paths of knowledge.

We will now look in more detail at a subject that inevitably arises as soon as we look for answers to the great questions of existence: the opposition between science and religion.

These two domains try to provide answers to the big questions. Unfortunately, there are so many inconsistencies between them that many believe they have to choose between these options. But, is it normal that these two important domains of human existence are irreconcilable?

To see more clearly, we will resort to a rarely used means: information theory. This theory may seem to have nothing to say about this conflict, but, on the contrary, it can give us simple solutions, as we will see.

Information theory is certainly one of the most important scientific achievements. Modern societies depend on this theory, which allows mathematicians, programmers and engineers to process information in ever more efficient ways. It has given birth to the computer, as well as the Internet, on which our economies rely. Just as we can talk about past eras by naming them "the Stone Age" or "the Iron Age," we can call our age "the Information Age."

Information always tends to circulate, and the way it does so is also explained by information theory. This process can be summarized by the figure next page.

This figure is simplified to the maximum. In reality, there are always many intermediaries between the transmitter and the receiver, which act as coders and decoders that do translation. They take a signal in a domain and transpose it in another domain, obeying the law of reproduction.

When we look in detail, even the simplest communication process

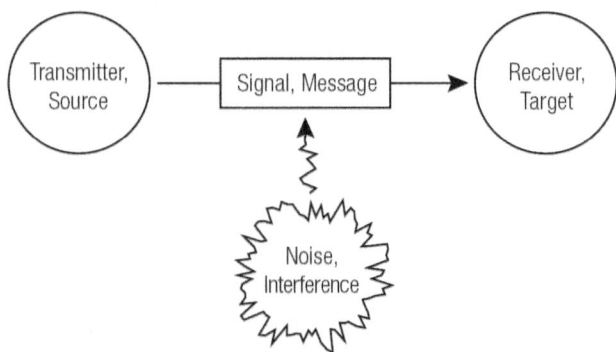

appears as a complex chain of transmission, where many translators are at work. For example, when someone speaks to us, there is a whole series of intermediaries doing translation. First of all, there is the brain of the person speaking to us who has to translate thoughts into words. The instructions to form these words are then sent, in the form of electrical signals, to the muscles of the mouth and the throat, which translate these signals themselves into muscular contractions. This way, impulses are transmitted to air that carries the signals to us in the form of sound waves. Our ears then translate them into other electrical signals that are sent to our brain via the nerves, and our brain finally decodes them to discern the meaning.

Whether it is translating from one language to another, the flow of information over the Internet, or any other process involving the transmission of information, there are always many steps of this kind.

Circulating this way, the information must continuously move from one medium to another and undergo a translation process each time. At each of these steps, errors can be introduced, errors that can decrease the quality of the signal, and even distort it completely. That is why, in the previous diagram, the notion of "noise" is illustrated to emphasize the fact that the message is continually threatened by external influences. Here, the notion of noise is used in a broad sense; it is not only irregular sounds coming from the environment, *but any irrelevant information,* that is to say, information that has no place in the signal.

The notion of noise is not trivial since much of the work of information theorists consists of finding solutions to this problem. This is a significant problem in our time, where the signals often travel thousands of kilometers, meeting on their way all kinds of obstacles that threaten to degrade them. This is not just a matter of concern for specialists because noise is something we continually struggle with in everyday life. For example, when we have to raise our voices to make ourselves heard in a noisy

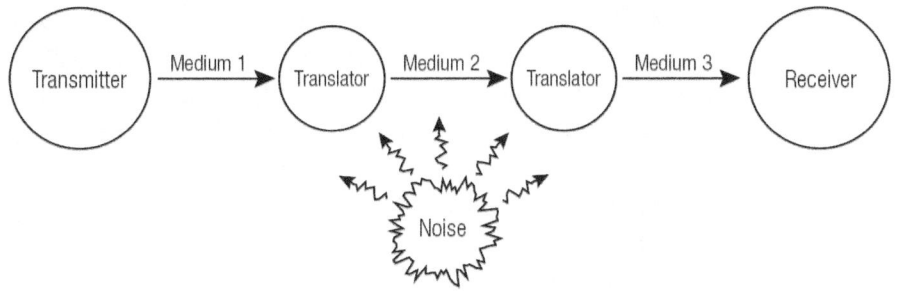

When transmitting information, the signal must always go through many translations. At each step, the signal may be contaminated by noise.

environment, when we have to say many times to our child that dinner is ready because he is distracted by something else, when an advertisement interrupts a video that we are watching, or when we cannot concentrate on our work because we are distracted by worries.

This was a summary of how information flows, and we will now see how it can help us in our reflection about science and religion. First of all, we must understand that what science and religion offer us is ultimately nothing but information. It does not matter whether we are talking about a scientific article or a religious text, what we are offered is information on a subject, even if this information can sometimes be of bad quality.

Therefore, the rules we saw previously must apply. This means that the information presented to us, regardless of whether it is scientific or religious, always had to go through many mediums and many translations, and most importantly, *it is always likely to have been contaminated by noise*. Because this is the most important work we have to do, regarding either a scientific or a religious discourse: to separate the relevant information from noise.

Here, many will find it odd that we treat religious discourses in the same way as scientific discourse, seeing the two simply as sources of information because, in our day, many people consider that science is the only one that can inform us about reality and see religions only as a set of superstitions, inherited from an age when humanity was not yet illuminated by the light of reason. Seeing religions only as a set of myths and dismissing them offhand is an easy solution used by many materialists. But adopting a blind skepticism toward religions is as false as adopting a blind faith because reality is more nuanced than that.

From prehistoric cultures to the present day, religion accompanies humanity in all kinds of forms. It is quite a *natural* practice, which comes from the fact that the human spirit carries with it the need to keep in touch with the invisible realms from which it comes, the sublime realities beyond material reality, and religious rituals can help us accomplish this. In addition to this need to reconnect with one's origin, everyone, unless completely corrupted, also has an impulse toward goodness, and the role of religion is to maintain that impulse, this attraction toward what is noble and ideal, this thirst for light...

Of course, this is *in principle* the role of religions, but one that too often they do not fulfill! This is because they have introduced in their discourse a tremendous amount of noise. So, it is here that nuances must be introduced: Religions should be beneficial, but they are now contaminated with all kinds of foreign elements, parasitic noises that confuse the messages that religions should transmit.

It is not necessary to exemplify with one religion in particular, for they are all as distorted as one another. The great religions are organizations much more concerned with earthly power than with the good of humanity. When given too much power over our lives, they often become a dictatorship that imposes its rules that come from arbitrary interpretations of ancient texts, rules that give disproportionate power to the religious authorities and their allies, and lock up the population in a mental prison, threatening all those who want to free themselves of severe consequences. History is filled with terrible examples of this behavior.

Materialists are quite right to be wary of religions because they do contain falsehoods. But they commit a grave error when they begin to believe that everything that the religions offer is false, because, behind the noise accumulated by the religions over the centuries, we find fundamental truths without which it is impossible to give meaning to existence.

Among these truths, there is the fact that the true nature of the human being is not material, that there are invisible worlds full of life and that the evolution of the visible worlds is guided from these invisible domains. These broad lines are found in almost all religions, but on this sound basis, they have added all kinds of inventions, parasitic noises, which, over time, ended up suffocating everything. The most harmful of these misconceptions is undoubtedly the belief that there are domains that are not subject to natural laws, that is, supernatural worlds. Adopting this posture, the religious authorities disregarded the laws and painted a portrait of reality devoid of any logic.

Therefore, the error of the materialists is to believe that religions have no knowledge to offer us, while we find great truths in the heart of religions if we manage to filter what is only noise. This knowledge has not been obtained by laborious research, as is often the case in science, but in another way, the path of revelations.

The idea that one can obtain knowledge by revelations is also something natural. To understand it, we only have to replace the word "revelation" with the term "inspiration." Indeed, history is filled with exceptional beings, who marked their time because they received great inspirations. This is both in the field of the arts and in the field of science, where we then speak of "genius ideas." The revelations in the heart of religions are just other examples of this process. It is a particularly high form of inspiration, which has allowed vital knowledge to be transmitted to humanity, through people serving as channels of transmission, as messengers.

Through our inner life, each of us is continuously interacting with the invisible worlds, and we have the ability to connect to the infinite variety of information that exists in these worlds, like some kind of Internet of nature. There is no limit to all that one can receive in this way, but because of our down-to-earth concerns, we remain too often connected to the lowest domains, the heaviest and most insignificant; while others succeed in connecting to very high domains, and what they receive in this way can transform them, and even change the world.

Once again, there is nothing magical in these processes, all of which respect the laws of nature, as well as the rules of circulation of information mentioned at the beginning of this chapter. The law of selection is particularly relevant here since it is the quality of our inner life that determines with which currents we can interact. These bonds cannot be forced, they are realized naturally when the right conditions are fulfilled; so, it is impossible for someone to become linked to the highest levels without the corresponding qualities. Therefore, to be able to receive inspirations from the highest planes, we must first work on ourselves to develop these qualities.

Inspired people then become *translators* because their role is to take the information they have received from the higher levels and to translate it into the material domain in the form of actions, words and works of all kinds. In the case of revelations intended to guide humanity spiritually, this translation has generally been in the form of words, which are a limited medium. To overcome this limitation, we can use *symbols* to convey the essence of a message through images and metaphors. This is why religious texts are full of symbolic narratives, stories that we must understand *intuitively*, and not interpret literally.

Whether we talk about spiritual revelation, artistic inspiration or genius idea, *the same process is always at work.* All that changes is the level where the inspiration comes from, that is, where is located the source of the information that the inspired person captured as a receiver. Also, in the same way that some artists claim to be very inspired when they are without originality, and some scientists think they are geniuses without really being, we must not believe all those who claim to receive revelations from the spiritual heights. We must remain vigilant because, most of the time, all we are offered is noise.

To see things more clearly, we must separate what is "revelation" from what is "religion." What religions offer us is never pure revelations, but interpretations of revelations that humanity has received in the past, translations that are often wrong. The proof that religions contain many errors lies in the fact that there is a great deal of contradiction between the different religious doctrines; therefore, they cannot all be true. A problem that religious fanatics overcome by believing that their interpretations are the only ones that are right, while the materialists see these contradictions as proof that there is nothing true in religions. In reality, this means that we all have a job to do to separate what is true revelation from what is only parasitic noise. The best tool for doing this work is the universality of the laws of nature: *Only that which respects natural laws can be true.* This is a criterion that everyone can apply, as much with the intuition as with the intellect.

Instead of talking about science and religion, it would be better to speak of science and revelation and to see the two as complementary, as two different ways of acquiring knowledge.

In reality, gaining knowledge by way of revelation is not exceptional because we continually receive information in this way in everyday life. If we consider the notion of revelations in its broadest sense, we can say that knowledge is revealed to us when it is not the fruit of our own research, that is, when we receive it *passively.*

This applies in the field of thoughts and ideas, but also in the visible domains. The most common example of this way of acquiring knowledge is the relationship between teachers and students, where knowledge is passed on to students without them needing to do all the research that led to these discoveries. Considered broadly, there are all kinds of ways to acquire knowledge by revelation: by reading a book, listening to a newscast, browsing the Internet and so on.

In the case of revelations coming from the world around us, the sources are at the same level as we are; whereas, in the case of spiritual revelations, the sources are in the invisible realms. That is the only difference because it is just different examples of the same process.

Unlike the path of revelation, which is a passive method of acquiring information, the path of science is an active method. When we are in the current of revelation, it is the unknown that comes to us, whereas when we are in the current of science, it is us who go toward the unknown. Scientists are people who do not want to wait for someone to reveal to them what they want to know, but who seek to acquire by themselves knowledge about the subjects that interest them. They want to see for themselves and test themselves.

To illustrate the difference between science and revelation, we just have to imagine an object hidden behind a veil. The object is unknown to us, and to discover what it is, there are two ways: Either someone raises the veil for us, or we lift it ourselves. In the first case, we have acquired the knowledge of the object by revelation, while in the second case, it is by our own means, like a scientist. Moreover, this notion is included in the word "discovering," which means "removing what covered."

Science and revelation are the two complementary paths of knowledge, and both are indispensable to get a complete picture of reality. Trying to get a complete picture of the world using only one of these two methods is like running a marathon jumping on one leg! There must be no opposition between science and revelation, nor must there be any conflict between the right leg and the left leg: the two must work together.

Here is a figure that illustrates the complementary aspect of science and revelation, those two currents of knowledge that circulate between the known and the unknown, the visible and the invisible.

To act as a scientist is to go by ourselves toward the unknown and the invisible. We can even see the history of science as a journey toward the unknown. Much of the work of scientists consists of discovering

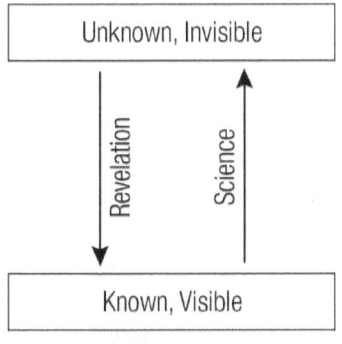

Science and revelation are the two complementary currents of information that flow between the known and the unknown, the visible and the invisible.

and classifying all the wonders contained in the invisible, that is to say, the previously unknown domains.

For example, the invention of the microscope allowed us to discover a rich and vibrant life, previously unknown because it is invisible to the naked eye. In another area, the telescope has allowed us to discover that the universe is much larger than previously thought and that it is filled with a wide variety of previously unseen celestial bodies. Similarly, physics has discovered many previously unknown particles, thanks to ever more sophisticated instruments. This rule can be applied to all sciences, since as soon as researchers succeed in taking another step in this adventure toward the unknown, what they discover goes beyond all the conceptions maintained by their predecessors. It is this reality that we have already summarized in this sentence: *Nature is much richer than we think*. The true richness of nature far exceeds what we can conceive, and so far, science has only raised a corner of the veil.

The adventure of science is something wonderful; however, like all adventures, it is not without danger. The main danger for scientists is pretension, intellectual pride. This is the monster threatening them at every turn of the road…

The danger of intellectual pride is to believe that our knowledge represents a "pinnacle," that our science is almost complete. It is enough to study the history of science a little, to see how easily intellectual elites fall into this trap. Regardless of the time, we always find elites who present their science as a peak of knowledge, just before new discoveries show us that their "great science" was actually very small!

A classic example is 19th-century physics, which many scientists of the time believed to be almost complete, just before quantum physics and general relativity revolutionized everything. The situation is similar today, where materialists believe that science knows enough to say that consciousness comes from the brain, and life from matter, while most is yet to be discovered on these topics. The main consequence of intellectual pride is that it closes us to everything that contradicts our favorite theories. It makes us like religious fanatics, who also believe they hold "the truth" and fight against everything that opposes it.

Throughout this book, we have repeated that it is a great mistake to construct our theories of consciousness and of life only on what is visible for the instruments of modern science. To escape from this error, the solution is to build our theories of consciousness and life only on fundamental laws, which will not change, no matter what will be discovered in the future. It is these laws that allow us to understand that the invisible part of reality is just as rich as its visible part and that the invisible contains the source of consciousness and life, as the great revelations always told us.

The way of science allows us to gain, by our own means, better knowledge of what was previously unknown and invisible; but on the other hand, it remains an enterprise that is limited by the technical capacities of a given time, and by the limits of the human intellect. There are realities far beyond anything that will ever be accessible to the instruments of science, and the great revelations have allowed humanity to have access to these realities, if only in a rudimentary form, through symbols. Spiritual revelation allows us to gain a broad knowledge about these areas, which allows us to understand life in general, while science, for its part, allows us to improve our everyday life. Revelation can give us an overview, including even the worlds that are invisible to us, while science gives us the details about our visible environment and allows us to master it.

These two ways of approaching reality are not in contradiction, *they are complementary,* like the right leg and the left leg. This appearance of contradiction results from the fact that the science presented to us is often interpreted in a purely materialistic way while, on the other hand, the revelations suffer from the questionable interpretations of the past. In other words, what is presented to us as useful information is often only noise.

As much on the side of materialism as of religions, one finds this same contempt for the laws of nature, this same necessity to rely on exceptions to the laws to allow false beliefs to subsist. Thus, materialistic and religious philosophies have much in common, since both are *parasites,* each in their respective domains. Materialism is a parasite of the domain of science, while religions are parasites of the domain of revelations. Both live *at the expense* of their respective fields, masquerading as beneficial elements, whereas they are only mental diseases.

Seeing things this way allows us to understand better the state of confusion in which humanity is lost regarding the great questions of existence. Scientists should be the guardians of science, while religious people should be the guardians of the revelations, but both have failed in their mission! Scientists have succumbed to materialism, that is, they have decided to give too much importance to the visible domain. While religious people decided to see the invisible as a supernatural domain, which they filled with figments of their imagination.

The solution is to rely on the invisible to solve the big questions, as the revelations always told us, but to consider the invisible as a domain subject to natural laws, these same laws that are at the heart of science. It is in this way that we can construct a *unified* vision of reality, in which science and revelation are no longer in conflict.

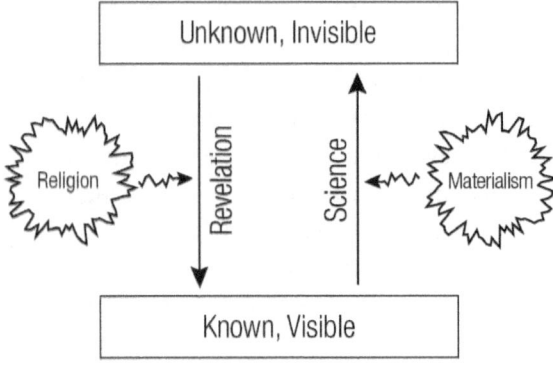

Religions are parasites of the current of the revelation, while materialists are parasites of the current of science.

11.1 INTELLECT AND INTUITION

Science and revelation can work together very well, provided that science is combined with true revelations, not fabrications. So how do we distinguish what really comes from revelation from what is superstition? Indeed, when there are doubts in the field of science, it is possible, using the scientific method, to test the ideas that are proposed to us. On the other hand, in the realm of revelation, information comes from sources that are inaccessible to most people, so how do we verify the veracity of what is said to us?

To see things more clearly, we again need to go back to the basics. In reality, humans have only two tools to verify the truth of an affirmation: intellect and intuition. On the side of the intellect, or of reason, what is needed is to *reflect* to see if what is presented to us possesses logical consistency, that is to say, is without contradictions, and to check if it agrees with the facts and laws known to science. It is this approach that we have used in this book, demonstrating that universalism possesses great logical consistency and that it fits very well with science.

On the side of intuition, it is more about *sensing* if what is presented to us is right. Therefore, it is a tool that works differently from the intellect. Intuition does not go into detail like the intellect because it works with synthesis, overview. On the other hand, it is able to instantly sense the value of a thing, or tell us if there is something wrong, even if it is not always able to tell us exactly why.

We all already have an idea of how intuition works because it is similar to the way we evaluate art. For example, in music, we all have the ability to recognize a harmonious chord just by listening. Similarly, in the visual arts, everyone finds that symmetrical forms are more harmonious than irregular forms without resorting to an intellectual analysis to make this judgment. Also, in the moral domain, everyone spontaneously finds that altruistic acts are more beautiful than egoistic acts. Examples like this could be given in all areas.

Intuition is what allows us to perceive beauty; one might even call intuition the "sense of beauty," which can also be seen as a sense of harmony, justice or balance.

At first glance, relying on our sense of beauty to make a judgment may seem unscientific, but it is enough to study the history of science to understand why the sense of beauty is a reliable guide. Indeed, when developing their theories, scientists naturally tend to prefer solutions that are *elegant*. Even if the sense of beauty is not precise enough to tell us exactly what is the right answer in every case, it is certainly a good guide to tell us what *direction* to follow.

This is especially true in the field of physics. Physicists see great beauty in the formulas of physics, and the notion of symmetry is central to the most influential theories of this field. The great geniuses of physics have always been in love with the pure and austere beauty found in mathematics, and in that sense, they have much in common with artists. Beauty is present everywhere in nature, so it is only logical that beauty is also present in the laws that are the basis of nature.

Beauty is a useful guide in science, and it is also a useful guide in the field of revelations. *Great truths must necessarily be beautiful,* and we can rely on this criterion to reject much of what religions offer us. For example, when one uses religious pretexts to justify discriminatory abuses, one can be certain that these justifications are false. People who are free in their judgment, without prejudice, clearly feel the moral ugliness of these behaviors.

Just as consistency and beauty are proper criteria for judging the truth in science, they are also essential criteria in the field of revelations, since the same criteria must be applied to *the whole* of reality. All the great truths must be beautiful and consistent.

Intuition is a reliable guide to help us move toward the truth, provided, of course, that it is really our intuition that speaks! Like all information sources, intuition is also likely to be threatened by noise, and the greatest source of noise that threatens the intuition is our *sentiments*.

Indeed, just like intuition, the sentiment is a form of sensation, and it is possible to confuse the two. The primary difference is that intuition is based on *universal* values, whereas our sentiment is based on *personal* values, namely our tastes, preferences and interests. For this reason, a person can pretend to "feel the truth" of a false belief when, in fact, he or she only feels a sentimental attachment to it.

The reasons that can lead us to attach ourselves to a false belief are many. It may be because this belief flatters our ego, it can also be the blind respect of traditions, the fear of being judged by our peers if we think differently, intellectual laziness and many other reasons. All of this can lead to intense feelings of attachment toward our false beliefs, so that we do not listen to our contrary intuitions anymore, through the noise generated by our emotions.

11.2 IN THE LIGHT OF TRUTH – THE GRAIL MESSAGE

The best way to move toward the truth is to focus on an overview that is as much in line with the basics of science as with the essence of the great revelations.

Materialists are quite right when they say that the supernatural does not exist, but they are mistaken when they believe that it means that the invisible worlds do not exist because the existence of invisible worlds is in accordance with the laws of nature; what is at odds with the laws is the distorted view that religions give us of these worlds.

Nor is there anything strange in the idea that we can receive messages, instructions or help from these invisible worlds, in the form of inspiration or revelation, since this can be explained with the same laws as any other information transmission.

Revelations can be seen as a rope offered to humanity from the invisible worlds that we can use to lift ourselves upward. The reaction of the materialists is to despise this help; whereas, the reaction of the religions is to seize the rope, but to use it to bind humanity and not to liberate it spiritually because religious authorities see the interpretation and control of these messages primarily as a tool for gaining power.

We add to this the fact that these revelations often date back many centuries or millennia, that they have reached us after having passed through innumerable intermediaries, accumulating errors over time, and we understand why religions are unreliable sources.

The solution is to look for a source that is not degraded by noise in this way. For this purpose, I recommend the work *In the Light of Truth – The Grail Message*.

Written in the first half of the 20th century by Abd-ru-shin, a German author, this work gives natural and logical answers to the great questions of existence. All the most important topics are covered: the true nature of humans, their origin and evolution, their place in existence, their responsibility, life after death, the invisible worlds and their inhabitants, and much more. We find there the many subjects that are at the heart of religious philosophies, but without the usual distortions.

I regard the Grail Message as the highest source of revelations. Of course, reading this statement, many readers will wonder what criteria I use as a basis for such an opinion. The answer to this question is very simple since it consists of the same universal criteria of truth that we saw earlier in this chapter, that is, *logical consistency* and *beauty*.

When I say that this work is the best source, it means that in my view it is the most consistent and the most beautiful. In reality, the two go together as two sides of the same coin, since consistency is always accompanied by beauty, and vice versa. The Grail Message has a pure and austere beauty similar to that of a transparent

"In the Light of Truth – The Grail Message" by Abd-ru-shin. A unique work that gives natural and logical answers to the great questions of existence. To know more, visit grailmessage.org.

crystal, because the author always follows a rigorous logic centered on the simplicity of the natural laws. And the portrait of reality that it present is the most beautiful that it is possible to conceive...

Universalism, the theory I built, is based on two pillars: science and the Grail Message. The purpose of my book is to present the scientific basis of universalism, showing how the laws that are at the heart of science can be used to solve the mysteries of consciousness and life. For those who want to know the other pillar, they must read this unique work that is the Grail Message.

To those who want to go further in their research, I highly recommend reading *In the Light of Truth – The Grail Message*. As for those who believe that science is the only way to acquire knowledge, it goes without saying that such a work is not addressed to them. That being said, whether or not one decides to value what comes from revelations does not change anything about the arguments that have been presented in my book since they are based on science. My conclusions are logical deductions based on laws that everyone can verify by themselves, *therefore, they do not depend on any revelation.*

If there are similarities between the ideas presented in my book and those found in some revelations, it is quite natural because there have never been contradictions between science and revelation; the two are only different ways of knowing the same reality. The separation that is usually presented to us is only an artificial division that materialists and religions each maintain in their own way.

Science and revelation are two different ways of progressing in our knowledge, either from the bottom up, as is the case with science, or from the top down, as is the case with revelations. Both methods are useful and necessary, we only need to learn how to use them properly.

12. WHAT IS A GOOD THEORY?

The ideal is to explain a lot with little.

The big answers are in the great laws—this is the essence of the message of this book.

In sum, universalism tells us that the answers are in the universality of the laws, the invisible and energy, and not in exceptions to the laws, the visible and matter. The proofs refuting materialism are not at the cutting edge of science, but in the most well-tested laws of science. When one approaches these questions without preconceived ideas, by accepting only the pure logic of the laws of nature, the mysteries disappear, to be replaced by answers of perfect clarity.

The questions concerning the true nature of consciousness and life are among the most important ones. Therefore, the answers to these questions must be found in the most important laws, which are also the simplest. *Everything important to understand is easy to understand!* Contrary to what materialists believe, these answers are not found in strange phenomena that have escaped science until now, but in the great laws that are the *basis* of science. Everyone already knows these laws intuitively, since we experience them at every moment. We just need to deepen them, going beyond appearances.

A saying that is often used to counter theories that move away from materialism is that "extraordinary claims require extraordinary evidence." According to this idea, to be successful, non-materialistic proposals would require spectacular evidence, which would shake up science. But, when we take a closer look, we realize that materialists have a rather strange conception of what is, or not, an extraordinary claim...

What is extraordinary is to assert that a material object can generate consciousness and that life can emerge spontaneously, because it goes against everything science knows about how reality works. Whereas, on the other hand, universalism considers that all material objects are unconscious and that all life forms are reproductions, assertions that cannot be more common since everyone normally considers things this way! Similarly, there is nothing extraordinary about placing the invisible and energy at the center of our conception of consciousness and life, since science has always confirmed that the invisible and energy are what is most important in nature.

There is no need for "extraordinary evidence" to solve these questions since the evidence is in the most ordinary laws. Like any theory, it is by testing universalism that one proves its value, and it is possible to test universalism simply by testing the laws on which it is based. Whenever an observation fits with these great laws, the universalist approach is validated, for it is these laws that prove that consciousness and life come from the invisible side of nature and from energy.

These proofs are indirect but are still strong evidence. Some think that indirect evidence is less valuable than direct evidence. In reality, there is no clear distinction between what constitutes "direct evidence" and "indirect evidence." It is always an arbitrary judgment. In fact, there are only different levels of indirect evidence, since there is always the presence of intermediaries between us and the elements we observe; even our senses are intermediaries.

One could even say that the more science progresses, the more the proofs it offers us are indirect, since the objects it studies are farther and farther away from what is accessible to our senses. For example, the observation of the Higgs boson, thanks to the Large Hadron Collider, is considered proof that this particle does exist. But, when we take a closer look, we quickly understand that physicists are far from having seen the Higgs boson directly. They have only made observations that indirectly prove that this particle exists. They have caused billions of collisions, analyzed billions of signals, through an incredibly complex succession of devices and programs, and finally, what they have seen are just results displayed on computer screens and not the Higgs boson itself. Why are these numbers and images displayed on screens considered proof? Because physicists have *confidence in the laws* that govern the operation of their devices because these laws have already been tested in all sorts of ways throughout the history of science.

The same is true for all scientific evidence. Ultimately, it is always because we have confidence in the laws that we can say that an observation is proof, and this confidence comes from the fact that these laws have been tested countless times, both in the laboratories and in everyday life. It is the same for universalism: We can trust its conclusions are valid because the laws on which these conclusions are based have already been tested countless times.

Despite this, many will think that universalism is not a true scientific theory, for all kinds of reasons. This is a subject that could be discussed for a long time, because there is no generally accepted definition of what exactly a theory needs to possess to be called scientific, there are different schools of thought in the philosophy of science. This is due to the fact that the line is not always clear cut, that it is not "all or nothing," that there are nuances. Science is a gradual process, scientific theories are not born perfectly formed, they are constantly evolving. Because of this, there are not only the category of "scientific theories" and the category of "unscientific theories," but rather a gradation with theories that are more scientific than others. The more a theory is based on many different facts, the more scientific it is, and the less it is based on facts, the less scientific it is. With, at the bottom of the scale, the theories that are antiscientific, that is, those that are contradicted by many facts.

Universalism is based a lot facts, the most important of which are the laws of nature, which are the most well-tested facts of science. Laws that are expressed mathematically through the formulas of physics. Therefore, I consider this to be a robust scientific theory, and I believe that it will become more and more scientific over time as science progresses.

It is also a theory that is refutable, since the observations on which it is based are potentially refutable. For that, it would have to be shown that some of the laws of universalism do not always necessarily apply in the domains where this theory asserts that they must always apply. This would be the case if materialists succeeded in proving that a material object can generate consciousness, or that life forms can emerge spontaneously.

Materialists believe that such things are possible, and they are free to keep looking for evidence. But the reality is that the universality of the laws completely refutes materialism, just as it refutes the worldview that religions offer us, in which exceptions to the natural laws are also permitted. Instead of the cult of appearances of materialism, the absurd superstitions of religions, and the indifference of the agnostics, *we must choose the perfect coherence of the laws of nature.*

Even if we kept repeating this message thousands of times, we would still be far from emphasizing the real importance of the universality of the laws of nature. It is only by continually repeating what is essential, with the insistence of a jackhammer, that it will be possible to break the thick layer of false beliefs that surround the so-called mysteries of life, an envelope formed by misinterpretations that have been accumulating for millennia. It is only through many repetitions that one can form new mental habits, and that is why, in this book, I have not hesitated to continually come back to what is really important.

In the next chapter, we will summarize the most important points of universalism. But before that, let us return to the notion of theory, to better understand why universalism is a good theory, even if, at first glance, it seems different from those we usually see in science.

A theory is a representation of reality elaborated by the intellect, one could also say that it is a reproduction. This reproduction is a translation of reality in the language of the intellect, which is logic. This process is essentially the same as the formation of a plan or diagram. The intellect is like the cartographer who studies the world to reproduce it on a map, and we call "theories" the representations that the intellect forms by establishing the logical relations between the elements we observe. We can also consider these representations as some sorts of simulations. In short, the intellect is a kind of navigation system, and theories are the mental maps that it uses to explain, predict and plan.

It is important to understand that the process of forming a theory works by *successive approximations*. When it approaches a new subject, the intellect begins by forming a first model, which consists only in the outlines. Then, as it acquires new information, the intellect improves its model, either by correcting the errors it contains or by making it ever more detailed. It is this process that is at the heart of the history of science, where the pioneers who entered a new field of knowledge began by laying the foundations, upon which their successors built ever more precise theories. This process is similar to that of an artist drawing a subject. He or she begins by brushing the main elements, and then makes the drawing ever more precise with many successive steps.

By putting a lot of work into it, the artist can make the reproduction ever more detailed, but despite the best efforts, it is impossible to translate all the richness of nature this way. We can only grasp certain aspects of it. Likewise, the intellect is a limited instrument, and it cannot represent everything in its theories. On the other hand, it can very well grasp the outlines, the main laws, and use these laws to form approximate

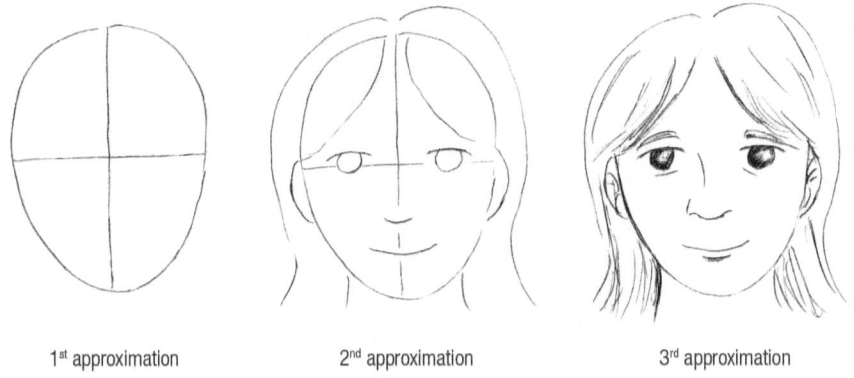

1st approximation   2nd approximation   3rd approximation

A theory is a model of reality, developed with many successive approximations. It is a process similar to the creation of a drawing, a plan or a diagram.

representations of reality, representations that do not need to be perfect, only to be sufficiently accurate not to mislead us. Just like a road map, which represents only what is necessary to efficiently navigate from one point to another.

One may attempt to criticize the worldview proposed by universalism, saying that this theory is not sufficiently precise in its answers for these to be considered true explanations. Among other things, it could be criticized for using some terms too broadly, such as the word "light," which is used to talk about all known and unknown forms of energy. It could be blamed for not specifying what substance the spirit is made of, and what types of particles or waves it uses to communicate with the body. It could also be criticized for not explaining precisely what are the different forms of invisible life, and not explaining exactly how this life has been transmitted to the visible side of reality, at the time of the origin of life on Earth.

This criticism comes from a poor understanding of what is a good theory. It must be very clear: *All theories are approximations.* To have value, these approximations do not need to be perfectly precise, only to be precise enough to be useful. The level of precision necessary for a theory to be useful depends on the *goals* that one pursues because it can be precise enough to be useful in one context, but insufficient in another context.

To illustrate this point, let us take as an example the different theories of gravity. Everyone, scientist or not, has an idea of how gravity works. Even prehistoric men and women had already understood the basics, and their theory of gravity could be summed up in one line: All heavy objects are attracted to the ground. This is a rudimentary theory, but it is still scientific in a certain sense since it comes from an observation of nature.

This rudimentary theory of gravity has always been with humankind, and it was not until Isaac Newton's work in the 17th century that there were major advances in the understanding of gravity. Newton's genius was to understand that objects are not only attracted to the ground, but that *all* objects attract each other. This attraction is proportional to the mass of the objects and inversely proportional to the square of the distance that separates them, data that can be expressed mathematically, which Newton did in his famous formula of universal gravitation.

Newton's theory made it possible to understand better how gravity works, but that does not mean that the prehistoric theory was bad. It remains a useful theory if we understand that it is applicable only in a certain context. Indeed, in everyday life, this rudimentary theory is quite correct, because the Earth is by far the most massive object in our environment, so all objects will be attracted by it. Therefore, the prehistoric theory is precise enough to allow us to navigate through everyday life, and it is only in special cases, such as when we try to predict the movement of the celestial bodies, that it is insufficient and we need to use the Newtonian theory. It is only a question of compatibility, the prehistoric theory is compatible with the aims of everyday life, and Newton's theory with those of astronomy. Therefore, both theories are good, provided they are used within the right context.

Newton's understanding was a great contribution to science, but even this theory remains an approximation that does not work in every case. The next step in the understanding of gravity was accomplished by Albert Einstein at the beginning of the 20th century, thanks to his theory of general relativity. In Einstein's formulas, gravitation is seen as an effect of the curvature of space-time, a revolutionary concept at the time. In the majority of cases, the results of the formulas of Einstein and Newton are almost the same, but in some special cases, the results are significantly different, and these are the cases that show us that Einstein's theory is more precise. The best-known example is that of the orbit of Mercury, which contains an anomaly inexplicable with Newtonian formulas, but that Einstein's formulas allow to predict with precision.

Therefore, Einstein allowed us to take another step toward an ever more precise understanding of gravity, but, again, that does not mean that Newton's theory was bad; it remains a good approximation that works in the majority of contexts. This allows us to illustrate a point that was mentioned earlier: Theories are elaborated by successive approximations. Thus, the prehistoric theory is the first approximation of the functioning of gravity, Newton's theory is the second, and that of Einstein, the third. Each time, a higher level of understanding is reached, but that does not mean that the previous approximations were bad, because they remain quite sufficient in the right context. In daily life, Einstein himself did not use his theory when thinking about gravity, but rather the prehistoric theory, because it is by far the most efficient; it allows us to make predictions that are almost always right, and with very little mental effort. When he was at the top of a ladder, Einstein did not begin to calculate the geometry of space-time to predict what would happen if he fell; he used the same theory as prehistoric men, and everyone will agree that this is the smartest option!

Another well-known example of a theory that has developed through successive approximations is the theory of the atom. The idea of the atom is very old. It comes from some Greek philosophers of antiquity, who conceived of atoms as small indivisible balls. That is why the word "atom" comes from Greek and means "indivisible."

This conception of the atom remained virtually unchanged, until Ernest Rutherford, at the beginning of the 20th century, demonstrated that the atom actually possessed a very dense nucleus, around which there are particles in orbits, the electrons. In this theory, the atom is no longer an indivisible ball, but a sphere that is essentially empty, like a miniature solar system composed of a central charge around which the electrons rotate like tiny planets.

This conception of the atom is still widespread today, but it is only a more precise approximation than that of the philosophers of antiquity. It is not exact since it does not take into account the discoveries of quantum mechanics. Nowadays, physicists no longer conceive of the atom as a small solar system, but as a complex structure, in which the electrons move around the nucleus in vibratory domains called "orbitals," which can take a wide variety of forms.

Despite all these discoveries, it is still common, even in scientific circles, to conceive of the atom only as a small ball, since it is often a sufficient approximation. For example, when chemists imagine atoms inside molecules as balls connected together by sticks, they know very well that it is not exactly right, but in those cases, it does not matter.

These examples allow us to understand better what a useful theory really is. A useful theory does not need to be perfectly precise. Such a thing will never exist, since all theories are approximations; like a two-dimensional image, which, despite all our efforts, will never be a perfectly accurate representation of its three-dimensional model. The best theories are not necessarily the most accurate ones, but the ones that are the most *effective* or the most *well adapted*, that is, those that allow us to understand easily what is most important within a particular context, avoiding unnecessary reflections.

Let us look again at the example of the road map. Indeed, if we give car drivers the choice between a road map and a satellite photo of the region shown by the map, all will prefer to guide themselves using the map, even if the photo is a much more accurate representation. Why? Because the road map already clearly shows us what it is important to know, while the photo obliges us to do some analytical work to separate the information that is essential from what is secondary. In other words, with the photo, we must do ourselves the work to develop a mental map containing only the relevant information, while with the road map, this work is already done! This is the most efficient option, the smartest one, the one that saves the most time and energy. In short, it is the best option, since it allows us to avoid unnecessary reflections.

We do not say that a road map is bad because it does not show us every house or tree along the way. On the contrary, a map that shows all these useless details would be considered a worse map than one that shows us only what is important. The same is true for architectural plans, pictograms on road signs, logos and so on. These are simplifications that focus on what is essential, and are appreciated precisely for that. It is the same for theories: The good ones are those that show us only what we need to know within a particular context, to answer the questions we are asking.

In the same way that good software gives us maximum results by using the least possible number of processes, to avoid that the computer unnecessarily spends its calculation time, a good theory gives us maximum results by using as few premises as possible to spare the intellect unnecessary reflections. That is why, despite all the advances in science, the prehistoric theory of gravity remains the most used, even if it is the least accurate, simply because it is most effective in the vast majority of cases!

This reflection allows us to understand better why universalism is a good theory, even if it is imprecise on specific points. It is even an excellent theory, provided that our goal is to understand what is *essential*, and not to explain everything in detail. In other words, if our goal is to answer the big questions, and not the small questions, that is a good theory. It allows us to answer the big questions, relying only on well-tested laws, and this by avoiding all kinds of unnecessary reflections, which put us at risk of getting lost in the details.

For example, concerning the substance of which the spirit is made, the universalist theory makes it possible to answer this question in broad outline, saying that this substance is in the family of light, and not in the family of matter. On the other hand, it does not tell us exactly what kind of light or energy, and it even says that many things remain to be discovered in this area.

Concerning the interaction between the spirit and the body, one may wonder what kinds of particles, waves or forces are used and whether they are part of the ones known to current science. Again, these are useless reflections, since we are only trying to form a first approximation, where such details are irrelevant. What is essential is to understand that the law of interactions is respected, that is to say, that this interaction is necessarily an exchange of energy carrying information. What is the medium of this exchange? This is a question to which the universalist approach does not respond exactly, but this is secondary since this approach works even if this question is not resolved.

It is the same for questions relating to invisible life. Precisely, what is this invisible life? This is a fascinating question, but one that can easily mislead us, so vast it is. We have already seen that the invisible life corresponds, in part, to what we call our "inner life," but it is only a small part of the answer. Invisible life works according to the same laws as visible life, which means that the invisible life is just as rich and complex as visible life, and therefore, to ask for a simple answer to this question is absurd. Millions and millions of species, combining their activity to form networks of infinite complexity, that is what life is—whether this life is visible to us or not does not change anything! We must not forget that this is only a question of point of view: What is visible for one creature is invisible for another. Each being is adapted to its field of activity and perceives only what is useful. This distinction between the visible and the invisible is arbitrary, since, from the point of view of the laws of nature, invisible life is like visible life. They are only *variants* of the same thing. That is what is important, and that is what the universalist approach is focused on. The rest is just details.

One might think that this attitude, which consists of leaving secondary questions unresolved in a theory, is exceptional, that it is not something that scientists usually do. But no, this is normal, especially when venturing into a little-explored area of knowledge.

Let us go back to Isaac Newton since he gives us a good example of this fact. When he developed his theory of universal gravitation, Newton realized that it worked very well to say how strongly objects attract each other. On the other hand, he admitted that his theory left a big question unanswered: It did not explain *how* objects attract each other!

Newton himself decided to leave this subject aside, and his attitude toward this question is summed up in one of his most famous quotations: "I frame no hypotheses" (Hypotheses non fingo). By this sentence, Newton admitted not having fully understood how gravity works, to only have understood certain aspects of it.

At first glance, this may seem very unscientific to develop a theory centered on the idea that all objects attract each other, without explaining how they attract each other! But, did that prevent this theory from being accepted, and this "bad" scientist from becoming a historical figure? No, simply because it is a theory that works, whether or not we understand how objects attract each other!

The example of Newton is no exception, for there are similar cases everywhere in science, which is quite normal since science has, above all, a practical purpose. We do not ask a theory to explain everything, only to reflect reality with *enough* precision to be useful within a specific context.

Moreover, the question of how objects attract each other, in Newton's theory, is the same type of question as those mentioned above, concerning the means of exchange between the spirit and the body, or between the invisible life and the visible life. Just like Newton, we do not need to answer these questions in detail for the universalist theory to be good. The important thing is that solutions that focus on the natural laws, the invisible and energy, work very well, even if we do not understand everything in detail. Because this is how science advances: We first try to establish which solutions work, not being limited by the fact that we do not understand everything from the start. This approach in science is not bad; on the contrary, it is the only way to progress!

Let us cite another example of this fact: The theory of evolution through natural selection, which had been accepted long before biologists understood how the mechanisms of heredity work, even if the notion of heredity has always been at the heart of the theory. Mechanisms of heredity that we are, moreover, still far from fully understanding, among other things because there remains a lot of discoveries to be made about how DNA works and the reasons which push it to transform.

Many other examples are also found in physics. In this field, researchers are continually working with abstract equations, without always understanding what they mean precisely in reality, and even having endless discussions about how they should be interpreted. This is particularly the case with the equations of quantum physics. These formulas speak of particles that are also waves (wave-particle duality), particles that are in several states at the same time until we observe them (superposition of states), particles the states of which are connected regardless of the distance between them (quantum entanglement), and so on.

These peculiarities produced by the formulas of quantum physics have been known for a long time; yet, there is still no consensus on the correct way of interpreting them. How do we separate what is real from what is produced because mathematical formulas and measures do not tell us everything? This question is so difficult in the case of quantum physics that many physicists prefer not to think about it. Their calculations are precise enough to be useful, and it is all they need; they leave the other reflections to the philosophers. This attitude is summed up in a humorous expression sometimes used by those who deal with quantum physics: "Shut up and calculate!" This expression is like a modern version of Newton's "I frame no hypotheses."

These expressions show us what is really important for scientific theories. Of course, this does not mean that a theory can say anything as long as it works because a good theory must always be anchored in reality. This means that a theory does not need to explain everything, only to explain enough.

Universalism is a theory that is perfectly clear about its goal: *to explain what is essential*. It is not a theory accumulating useless details and sophisticated concepts; on the contrary, it is a theory that seeks to be as simple and natural as possible because its goal is to be understood by the greatest number, while being rigorously built on the most important pillars of science.

The theory presented in this book is a first approximation, which is very precise at the level of the laws, but imperfect at the level of the details. This is because it focuses on the outline, on the big picture. All the phenomena described by universalism are extremely complex in reality, like everything in nature. But these secondary details must be left aside if one wants to see the great laws that act behind these phenomena. To understand these laws is to understand the functioning of reality, and thus, understand the core of what we need to know to answer the big questions of existence.

13. SYNTHESIS

Bigger is the question, simpler is the answer.

To look for answers in the great laws, not the small details, to look for answers in synthesis, not analysis, this is the essence of the universalist approach.

The laws of nature are the laws of physics, of energy, of movement, of life, of reality, of existence... The natural laws are universal and immutable; they will never change, no matter what will be discovered in the future. This is why they are the best foundation upon which to build a theory. In all their research, scientists never discover anything but different consequences of the fundamental laws, and nothing that contradicts them. Each discovery of science only confirms what everyone already knows intuitively: *reality is perfectly coherent.* Whether in the past or the future, here or there, in the visible or the invisible, the basics of reality are always the same. Existence as a whole is perfectly self-similar, or holosimilar. Simply because reality is formed *by* the laws of nature, and therefore nothing exists that does not conform to the laws.

The key that opens all doors is the universality of the laws of nature. We simply must stop believing that exceptions to natural laws are possible. This book was intended to explain how universalism can provide answers to fundamental questions and why this approach is entirely consistent with science. This is based on many reasons that all are invited to put to the test, to see for themselves their validity. The primary reasons are summarized in the following pages.

13.1 THE THREE PILLARS OF UNIVERSALISM

THE UNIVERSALITY OF THE LAWS OF NATURE

The laws of nature act the same way through all of reality.

The laws are universal principles, so they allow no exception in their field of application. If a law applies to an element of a particular kind, it must apply to all elements of the same kind. No matter where you are in the universe, in the past or the future, in the visible or the invisible, the activity of the laws is always the same.

Therefore, all philosophies based on exceptions to the laws are false. This is the case for materialism, since believing that matter can become conscious in certain special cases, or spontaneously engender life, is believing that exceptions to natural laws are possible.

THE INVISIBLE

Most of reality consists of phenomena that escape our senses, as well as the instruments of science.

Throughout the history of science, many enigmas have been solved thanks to a better understanding of the invisible. It is the same for the mysteries of consciousness and the origin of life, the keys to solving these questions are in the invisible.

Although it is not possible to have direct access to the invisible domains, we are still able to understand some essential aspects of them, thanks to the universality of the laws of nature. This means that the invisible worlds are similar to the visible worlds, that they are just as rich and alive since they are formed by the same fundamental laws.

When one accepts that consciousness comes from the invisible side of reality, the enigma of the brain disappears, since it is then seen as an object just as unconscious as the others, serving only as an intermediary for a conscious activity.

When one accepts that life comes from the invisible side of reality, the origin of life on Earth can be seen as a transmission coming from previous life forms. This transmission is a natural process following the same laws as other reproductions, the main difference being that invisible elements are involved.

ENERGY

The true essence of reality is energy, not matter.

Energy plays a central role in the greatest theories of science, and it must also play a central role in theories that attempt to explain consciousness and the origin of life.

Since we only interact with the world through energy, light in all its forms, it is logical that the center of our consciousness, our spirit, is itself made of a kind of light. The source of consciousness and willpower lies in energy, and we only see their effects in matter.

According to the principle of the universality of the laws, it must be the same for life. Essentially, it resides in energy, and its manifestations, in visible and invisible matter, are only secondary ramifications.

13.2 THE MAIN LAWS OF UNIVERSALISM

THE LAW OF EQUILIBRIUM

Everything tends toward equilibrium.

Everything tends toward the most balanced state possible, toward symmetry, uniformity, justice, harmony, unison. This is because energy always tends to be distributed as evenly as possible. Energy continuously seeks to produce a state of perfect equilibrium, and the universe is driven by this process that keeps it endlessly in motion since perfect equilibrium is unreachable in nature.

Like the other great laws on which universalism is based, this law is present throughout all sciences, under several names. For example, in the field of movement, this law manifests itself through the principle of least action; in the domain of mixtures, through entropy; in the field of statistics, through the law of large numbers; in the field of chemistry, through the principle of chemical equilibrium; in the field of biology, through homeostatic equilibrium; and so on. This law is also related to the law of retroaction since a retroaction always seeks to restore balance.

This law refutes the materialistic conception of the origin of life since it tells us that the spontaneous emergence of life is impossible because this imaginary phenomenon goes against the law of equilibrium, which constantly pushes all mixtures toward the most uniform state possible.

The negligence of this law is also the primary source of the problems that overwhelm humanity, as well as the personal level, as the social or environmental level, because we have broken some essential balances.

THE LAW OF RETROACTION

Every action comes with a retroaction.

This law is also called the law of cause and effect, the law of reciprocity or the law of compensation. In the field of physics, it is known as the law of reciprocal actions, which tells us that every action comes with an equal and opposite reaction.

This law gives rise everywhere to complementary currents, such as arterial and venous circulation, the motor and sensory currents of the nervous system, or inspiration and expiration. This law is also present in the living process of sowing and harvesting.

In the realm of human consciousness, this law manifests itself through the complementary currents of the will and of consciousness, that is to say, the current that comes from the spirit and the one that returns to it. This law also applies to the relationship between the invisible life and the visible life, where there are also complimentary currents, invisible currents that are at the origin of life on Earth.

THE LAW OF SELECTION

Every interaction is selective.

An interaction can occur only when certain conditions are met, conditions that vary according to the type of interaction. Similar to a system of lock and key, this law draws a line between the interactions that are possible and those that are not.

In the field of elementary particles, this law tells us that particles interact selectively according to the forces to which they are sensitive. When interaction is allowed, particles can be attracted or repelled. Otherwise, they are invisible to each other.

This law explains why invisible worlds exist. It also explains why humans, who have invisible parts, can interact with realities to which the instruments of science have no access.

THE LAW OF INTERACTIONS

Every interaction is an exchange of energy that carries information.

All exchanges take place through energy, light in its various visible and invisible forms. Therefore, every interaction can be described as a transfer of energy that carries information.

This is also the case for the exchanges between the spirit and the body, which take place through the currents of the will and of consciousness, as well as for the exchanges between the invisible life and the visible life, at the origin of life on Earth.

THE LAW OF INERTIA

Every form of matter only resists change.

This law can also be called the law of matter, since inertia, resistance to change, is the very essence of matter.

All that matter does involve resisting change; that is to say, it always keeps the same movement, as long as external influences do not force it to behave differently. This is why particles of matter can be considered "particles of inertia" or "points of resistance"; while particles of light, on the other hand, are particles of energy, which push all things toward perpetual change.

This law tells us that matter can never be conscious since it can never do anything but be inert. This means that the keys to understanding consciousness are not in the brain, or in the material domain, but in the energetic domain.

THE LAW OF REPRODUCTION

Every life form is a reproduction.

The law of reproduction is the central law of biology: Every life form is a reproduction, with variations, of a previous life.

The various beliefs in the spontaneous generation of life have all been refuted when it was understood that the life that was believed to be spontaneous actually came from previously unknown sources. The same is true of materialistic beliefs about the origin of life. They will be rejected when it will be understood that life on Earth comes from other life forms living on the invisible side of reality, an invisible life that functions according to the same laws as visible life, but that exists in domains of nature not well understood by current science.

The law of reproduction can also be defined in this other way: Every form is the reproduction of a model. The original model, the perfect model of which everything is an imperfect reproduction, is located at the point of origin of existence. We give it several names: God, the unique force, Life itself...

13.3 THE MAIN UNIFICATIONS OF UNIVERSALISM

CONSCIOUSNESS

The brain is unconscious, like all material objects.

The center of consciousness and willpower is the spirit.

The spirit is an invisible phenomenon, like most phenomena.

The spirit is an energetic phenomenon, like most phenomena.

The interaction between the body and the spirit works according to the same laws as the other interactions.

The spirit can perceive through energy because it is itself made of energy.

LIFE

The first life forms were reproductions, like all life forms.

Visible life comes from a transmission from invisible life.

Invisible life works according to the same laws as visible life.

There are life forms in the invisible domains, as there are in the visible domains.

There are life forms in energy, as there are in matter.

The origin of life and the origin of existence are one.

Life is the unique force at the origin of existence.

13.4 THE MAIN SYMBOL OF UNIVERSALISM

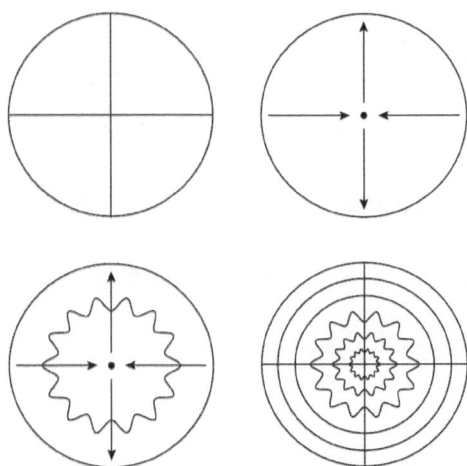

Universalism is fully summarized by the symbol of the cross in the circle, which can be presented in various ways. It is a diagram that shows reality as a whole, both in its structure and in its functioning.

Reality consists of visible and invisible levels, which can be material or energetic, formed by the law of selection. These levels are interconnected by complementary currents, governed by the law of retroaction. These currents reproduce information from one level to another, a process governed by the law of reproduction. These exchanges are intended to maintain the harmony, the order, the coherence, the health of the universal organism, governed by the law of equilibrium.

The levels are symbolized by the circles, the complementary currents by the branches of the cross, the processes of reproduction by the similarity of the circles, and the equilibrium by the equality of the branches and the symmetry of the circles.

This symbol represents reality as a whole, as well as each of its parts since they are only reproductions of the whole. As for the human being, the center represents the spirit, the cross represents the currents of the will and of consciousness, and the circles represent the different visible and invisible bodies of the human being.

In the general view, the center represents God, the unique force, Life itself; the cross represents the complementary currents that traverse the whole, produced by the laws of nature, laws that are the action of the unique force; and the circles represent the different levels of reality formed by the laws.

14. THE MEANING OF LIFE

The meaning of life is to serve.

In the eyes of many, claiming to know the answers to life's great mysteries is incredibly pretentious. There is a great irony in this attitude because it is precisely pretension that is the leading cause of the confusion surrounding these questions. These answers certainly do not constitute a "great knowledge" of which one can be proud because they could not be simpler, and we continually have them before our eyes in the activity of the laws of nature. The problem is that we are looking for solutions to these big questions hoping to find extraordinary answers, worthy of a "genius" or an "initiate," but since the real answers are only in natural simplicity, and even accessible to children, we refuse to recognize them.

Materialistic intellectuals do not value simplicity. For them, if the big questions are not yet solved, it is because they are very complicated problems. They are lost in a labyrinth of details without seeing that everything is animated by the same simple principles, those laws that we only need to study to obtain an overview that includes *everything*, even consciousness and the origin of life! The so-called mysteries of life are pure products of our imagination, and since these enigmas do not exist in reality, the only way to solve them is to get rid of the false beliefs that artificially created those mysteries.

It is the same with the meaning of life. It is not mysterious, and the confusion that surrounds this subject is also artificial, generated by those who do not want to content themselves with simple answers. Indeed, it is enough to reflect in general on what is capable of giving meaning, a reason for being, to obtain an obvious answer!

What gives meaning to the existence of a being or an object? The answer is clear: An existence makes sense only if it plays a useful role, if it serves a purpose, if it meets a need. It is the same for humans; to make sense, our life must be useful, so the answer to the question of the meaning of life can be summed up in two words: *to serve!*

The answer to this question is only common sense, and intuitively, everyone already knows it since everyone clearly feels that their life acquires value only if it serves something. To seek to be useful to others, to collaborate to collective happiness, to promote the evolution of the whole...these are different ways of expressing how we must behave so that our existence has meaning. Of course, the answer to the question "How to serve?" is different for everyone, according to our possibilities and our talents, but the principle is always the same: It is to have a beneficial effect on those who are in contact with us, improving our environment by our work and our way of being. The purpose of our life must be to improve the lives of others! In other words, it is to learn to truly love other creatures, since to love is *to want to serve.*

Morality is not a human invention because the basis of morality is found in the laws of nature, in reciprocity and interdependence. By observing the struggle for survival in nature, many people deduce that this is the basic principle of life and use it to justify their selfish behavior. They say they are following the example of nature when they exploit the weaknesses of others to achieve their goals, but their reasoning is false because life is actually based on mutual service and cooperation, not competition. Even a healthy competition is a form of mutual service, since it promotes the evolution of the parties involved in it; which is the case in the struggle for survival in nature, but is not the case when we abuse the power we have over others.

Everything is interdependent in nature, and this principle of reciprocity is what is most important because it is at the core of the functioning of the organisms themselves. In an organism, each part is at the service of the whole, and the whole is at the service of each part. The heart does not just beat for itself, the lungs do not just breathe for themselves and so on. Each organ sends the fruits of its work to the other organs and receives, in return, all that is necessary for its well-being. This reciprocity, which is anchored in the law of equilibrium, is the only thing that can ensure the survival of an organism, as well as the survival of an ecosystem or society; while the "every man for himself" mentality only brings about imbalances and divisions that ultimately lead to conflict, death and decay.

For an existence to have meaning, it must serve something. This answer applies to the life of a human being, and it is equally valid for the whole of existence since everything obeys the same fundamental laws. Humans have always looked up to the sky, wondering what the meaning of life is. Why does the Earth exist? The Sun? The stars? The universe?... We like to overthink, and we get lost in all kinds of intellectual subtleties, and in doing so, we forget that there is only one thing that can justify an existence: *utility*. So, to give meaning to the existence of the universe, we must first admit that it is useful, as well as all that it contains. The Earth exists because it is useful, as well as the Sun and the stars. The universe as a whole exists because it is useful, because it serves a purpose, it meets a need... The answer cannot be anything else since usefulness is the *only thing* that can give meaning!

To find answers to these questions, we must again use the principle of the universality of the laws of nature because all the answers are found there. Life is universal. Everything is part of the same living whole, and we must see the universe itself as an organism, an organism overflowing with living activities, most of which are invisible to us. As mentioned earlier, each part of an organism is at the service of the whole, and the whole is at the service of each part, and this principle of reciprocity must also apply to the cosmos. To give meaning to the existence of the universe, we must, therefore, consider that it is useful to the creatures who inhabit it, that it serves their evolution; and, reciprocally, that the creatures who inhabit the universe are destined to be useful to it, to serve its evolution.

As soon as we consider ourselves part of a universal organism, the answer to the question of the meaning of existence is self-evident: Humanity is destined to play a useful role inside this wonderful living whole. Just as within an organism all the elements have a useful function, humanity is itself an element of the universal organism that must justify its existence by being useful.

The laws of nature are *perfect*. They answer perfectly all the fundamental questions of existence because they are the basis of existence! These laws clearly show us that life can only acquire meaning through service. The alternative is to consider that existence has no meaning, that the universe is only a succession of senseless accidents, a chaos inside which life seems to be only a strange anomaly. This is the interpretation that is preferred by the materialists, and they dive into this abyss of absurdity, claiming that their "knowledge" does not offer them alternatives. But this "knowledge" is limited to appearances, to which they have decided to give more importance than all the rest, even if that implies ridding the world of all meaning, not listening to our deepest intuitions.

One of the fundamental errors of materialism is to believe that life is a strange exception, whereas the only thing we need to explain everything is to make life the fundamental principle of existence! Reality then appears to us as different links in an endless chain of life, links that can take an infinite variety of visible and invisible forms. Thus, there are no more mysteries, there is only life! A life that offers us marvelous possibilities, if we accept to insert ourselves harmoniously into this great chain, which is governed by the law of equilibrium: *We must give as much as we receive.*

Life offers humanity possibilities that go beyond anything we can conceive of, but to be entitled to it, we must respect the basic rule of life, which is reciprocity... Unfortunately, it is not difficult to conceive how much humanity has failed on this point! Indeed, the notion of a universal organism comes with another inevitable logical implication: In its current state, humanity must be considered a disease disturbing the harmony of this vast living whole.

All are able to recognize that this is only an objective remark: Humanity is not integrated harmoniously within the living environment of which it is a part. This is the very definition of a disease, and it is an abnormal situation that cannot continue for long without having terrible consequences.

With the crises that erupt in many domains, everyone is able to recognize that the current situation is critical. But we are always quick to blame others, holding them responsible for all evil. The reality is that we all share some responsibility for this lamentable situation. Of course, it is a speech that we continually hear these days, especially in the field of ecology. We will not repeat here those speeches that we all have heard a thousand times, but rather approach the question from another angle, emphasizing one aspect of the problem that is often neglected, while it is the most important: the fundamental role that plays our inner life.

In our time, it is generally considered that our inner life is not real and that thoughts have no consequences. We see this area as a playground where we do not need to control ourselves and where we can follow our darkest inclinations. This is because many believe those who tell us that our thoughts are just a product of our brain.

Throughout this book, we have insisted on the fact that most of reality is invisible. It is the same for the problems that overwhelm humanity today: Most of them originate from the invisible side of reality, in which we are inserted and with which we continuously interact without even knowing it. Everything is counted by the implacable laws of nature: every will, thought, desire, word, gesture... Our beliefs, our inner life and our way of being trigger invisible—but real—forces that form our reality.

For many, the idea that thoughts can be real forces is so terrible that they prefer not to believe it. Indeed, by noting the lack of control that exists in this area, one can easily imagine the flood of dark and harmful thoughts that are continuously pouring onto the invisible side of reality, giving rise to a form of pollution much worse than that which is denounced by environmentalists!

Before the rise of environmental awareness, the industries poured without any restraint all kinds of pollutants into the environment, believing that it had no consequences. Whereas, in fact, our wastes do not disappear magically when we stop thinking about them, but accumulate in our environment to the point that they get intoxicating. In the same way, to act with carelessness in the domain of thoughts is like urinating in a pool where you are swimming... It is a disgusting image, but one that symbolizes very well what has become of our invisible environment, in which we continuously pour out all kinds of filth that accumulate around us and act retroactively on us, to the point of intoxication. The clear water that could be our invisible environment has become, over time, an opaque and disgusting mixture.

Many receive ideas of this kind with much skepticism, but it does not matter if one considers our inner life as being real or not, *we still have to admit that everything starts from there.* Even if one refuses to believe that thoughts are real, it is stupid to believe that thoughts have no consequences because it is obvious that we must first *think* all that we realize in the material world, and therefore, that our actions are themselves only consequences of our thoughts and our will. In other words, *our outer world is only a reflection of our inner world,* and therefore, the first step for a real external change must inevitably be a change within us. We can lament as much as we want about the ugliness of the present world, but it is only the faithful reflection of what humanity is internally—it is only a mirror that shows us how we actually are.

If we want sustainable change, we must first focus on improving our thoughts, working to make them less selfish, purer, more beautiful, brighter. This summarizes all the solutions since, if we make our inner life better, our outer life will *automatically* follow. This cannot be done overnight, it is the work of a lifetime, but there is absolutely no other way to obtain a lasting result. All the other solutions only bring superficial and temporary improvements. By acting this way, *we act on the source of our reality*, and with time the situation also improves in the external world since it is only the reflection of our inner world, another manifestation of the law of reproduction.

By seeing things in this way, we eliminate another artificial division that plagues humanity today: the division that exists between the outer life and the inner life. Indeed, most humans try to control themselves outwardly because they know that this can have harmful consequences if they do not. On the other hand, in their inner life, many allow themselves anything, believing that in this area, there are no consequences for themselves and their environment. It is this division that must be eliminated by accepting that these two domains are equally real. In both worlds, our way of being brings consequences for which we are responsible.

Everything that is nourished internally for a long time inevitably seeks to express itself externally one way or another. This fact is the source of many evils, such as mental disturbances, harmful habits or uncontrollable impulses. The basic rule is very simple: *Everything that we nourish internally exerts on us a pressure that pushes us to express it externally.* It is perfectly natural, and by observing oneself, everyone can see that it is so. If one feeds a type of thought or desire for a long time, it acquires great strength and can become difficult to control. After committing a harmful or criminal action, we sometimes hear the person who perpetrated it say, "It was stronger than me." But in every case, these are only the ultimate consequences of a long process that would have been easy to stop at the beginning and now require great effort, just as it is easy to uproot a young shoot, but it is much harder to do so when it has become a tree.

To say that we must first focus on improving the quality of our inner life is simple and natural, to the point where some will be inclined to say that it is a naive solution to the problems of humanity. However, people who neglect the importance of the inner life see the world upside down. On the contrary, it is naive to believe that it is possible for humankind to develop harmoniously without addressing the root of all problems, which is the extreme negligence we display at that level. Many people will think that this is not really a solution since we cannot force anyone to change in this domain. Of course, we cannot force anyone, but we can emphasize the importance of this issue, showing objectively why we all *need* to improve our inner lives.

To solve the problems at the source, we must first work to improve the quality of our thoughts! We must make our thoughts purer, more positive, more luminous. By nourishing beauty within us, we become inhabited by a pressure that constantly pushes us to reproduce this beauty outside of us, by our words and our actions, so that with time, our environment becomes ever more harmonious and pleasant to live in.

We have the duty to keep our thoughts as pure as possible. It is a rule of life that is found in many philosophies, but that is too often neglected. This is because the consequences of our negligence in this domain take time to manifest in a visible way, which gives us the illusion that it has no real consequences. In the domain of good manners, we make a mistake by limiting ourselves to appearances. It is why civilization is often considered a "veneer," because it is only a thin outer layer that always ends up cracking under the pressure coming from the inside since we continually play with invisible forces without understanding their power. Forces that always end up manifesting themselves externally when they are nourished long enough because that is inevitable in the natural order of events.

That this great power has been entrusted to humanity is proof that it is destined to achieve great things, but this power can only turn against us if we do not use it properly. Just as a frail little shoot can become a great tree if it is fed for a long time, the seemingly insurmountable crises of our time are only the ultimate consequences of bad choices that seemed harmless at first, but the consequences of which have developed to produce bitter fruits in the form of crises in all domains.

To be stupid is to be slow to learn from one's mistakes, and one can say that the greatest catastrophe ever to hit the Earth is humanity's stupidity! The pretension of humanity is the stuff of legends, and it seems determined to follow its false ways to the end… For many, the current situation seems insoluble. Of course, if we try to uproot a tree with bare hands, we will not succeed and we will conclude that it is stronger than we are. But we must not forget that there is another way to overcome our problems: to stop feeding them. No one can bring down these monsters that we have created by confronting them directly. We must first weaken them by ceasing to give them our energy, and by channeling this same energy toward the solutions, which are like neglected little shoots.

The only lasting solution is to build something better, starting from the base: our inner life. There is absolutely no other way to improve the situation in the long term if we do not change how we are internally, by changing our beliefs, improving our thoughts and our way of being. It is not new technologies or new economic and social structures that will allow long-term improvements. All this has already been tried in the past, and humanity has always encountered the same problems in different forms because the root of all the problems—*the quality of our inner life*—has never changed! There are always the same fears, the same greed, the same selfish desires, the same stupid pride, the same uncontrolled animal impulses, the same ignorance of our place in existence…all this had manifested itself and still manifests itself, only the forms change according to cultures and times! So, those who live in an illusion are those who hope for improvement by neglecting the fundamental role of our inner life. It is impossible because our external world will always be only a reproduction of our inner world…

In summary, we can clearly see how the notion of a universal organism can give us an answer to the question of the meaning of life. We are an integral part of a living whole, which we must serve by playing a beneficial role, not only by our external actions but also by our inner life, which is even more important.

This vision is another vertiginous inversion from the materialistic view of existence: Instead of a sterile universe in which life is seen as an anomaly, we are part of a vast living organism *in which it is the current state of humanity which is abnormal.* It is not life that is mysterious, it is the ignorance of our place in existence that is abnormal! Just as is our ignorance about the origin of life and the true nature of the human being. This state of ignorance should never have existed because the mysteries surrounding these questions are purely artificial.

Humans are elements of the universal organism, among an infinite variety of others. Like all the rest, they must justify their existence by being useful, by being at the service of the great universal organism that gave them birth and on which they depend entirely. If they decide instead to behave in a harmful way, they become a disease, and if they do not change their way of being in the long run, they condemn themselves to be eliminated to preserve the health of the whole.

To eliminate undesirable elements the cosmic organism does not need to take any particular action, since there is an infallible self-correcting mechanism that takes care of it: the law of retroaction. This is because the undesirable elements condemn themselves to disappear under the retroactive effects of their behavior, which are amplified over time if they refuse to adapt.

These processes, which give rise to what is commonly called fate or karma, span millennia, creating the illusion that there is no universal justice. This illusion also stems from the fact that it is not possible to understand the functioning of universal justice without also accepting the notion of reincarnation, another idea against which there is much skepticism, but which is just another logical consequence of the universality of the laws.

The notions of karma and reincarnation are what we get when we accept that the law of retroaction is universal, and therefore, that it applies to *every movement,* which includes our will, our thoughts, our desires, our words, our actions... Every movement of our inner and outer life is followed by a retroaction of similar nature in proportion to the intensity of this movement, and all these retroactive effects combine to form our destiny. These retroactive effects are not directed toward the body, but toward the spirit, which is their true source, and when these effects cannot unravel entirely in one's life, there continue to be links between the spirit and matter, which can force the human being to reincarnate.

Once again, the basis of these processes cannot be simpler, it is entirely summed up in this well-known expression: *We reap what we sow.* In the same way that it takes time for a seed to develop, mature and give fruit, it takes time for universal justice to fulfill itself. This delay may extend for centuries or millennia, which seems enormous from the human point of view, but from the cosmic point of view, where time extends over billions of years, these processes can be seen as almost instantaneous.

The universality of the law of retroaction to which are attached the notions of universal justice, karma and reincarnation, is essential to give meaning to existence; and the proof of this is the state of confusion that is generated when one tries to understand life without these notions.

For example, many materialists see the inequalities that exist between birth conditions as a reason not to believe in the existence of God. Indeed, the great monotheistic religions support the existence of a *just* God, but how to reconcile this belief with the fact that some can be born in a privileged environment while others lack everything? The followers of these religions, who claim to believe in a righteous God, but who also think that we have only one earthly existence, face enormous difficulties in trying to reconcile these two beliefs. Their only response to these apparent injustices is the following: "God works in mysterious ways."

Yet, this is a legitimate question since we clearly sense that there is an inconsistency between this idea of a just God and the fact that there may be so many inequalities between birth conditions. These ideas are irreconcilable and lead us to a logical impasse. Materialists settle the question by abandoning the belief in the existence of God, in other words, by seeing the inequalities between destinies as proof that God does not exist. While on the side of religious people who do not believe in reincarnation, they decide to keep these two irreconcilable beliefs and maintain the artificial mystery, believing that the ways of God follow a strange and inaccessible logic.

These two approaches are false because the solution is to maintain the belief in a just God but to abandon the idea that there are injustices between birth. This can only be done by accepting that the situation in which we are born is a *consequence of our choices*.

So far, to describe God, we have used words like "perfection," "coherence" or "equilibrium," but we can also use the words "justice" and "love," as religions do. In reality, these are just different ways of talking about Life, the unique force. The perfect cause that is the basis of existence absolutely cannot allow inequalities without a good reason. This reason is not an impenetrable mystery: Each of us has an existence whose beginning far precedes our present life on Earth, and the situation in which we are born is the consequence of innumerable choices, a consequence determined by the laws of nature.

This does not mean that if we are born in a privileged environment, it is a sign that we have made good choices in our distant past, no more than being born in a difficult environment means that we have made bad choices. There are no rewards and punishments for the laws of nature, only logical consequences. Once again, one must be wary of appearances and avoid superficial interpretations, because, in reality, each situation presents different possibilities for the evolution of the spirit, as well as dangers. For example, being born into a wealthy family and growing up without material worries may seem like a good thing, but with material abundance, a person is at risk of becoming lazy and capricious; in the same way that water is a good thing for a plant, but too much of it can cause it to rot. Conversely, being born in an environment where life is difficult may seem like a bad thing, but the fact of being continually obliged to commit all our strength to ensure our well-being and that of our loved ones is favorable to the evolution of the spirit. The spirit can become accomplished only through service, whereas the excessive accumulation of material goods brings nothing and can even do harm.

Faced with questions of this kind, the most important is always to use natural logic. Religious people say they believe in the existence of the spirit, but many of them believe that the spirit can only be incarnated once on Earth. Yet, they know very well that if a phenomenon has occurred once, it logically means that it can happen again if similar conditions are met. This is a perfectly banal statement, a reasoning that everyone follows spontaneously in everyday life. So why do these people refuse to follow this reasoning concerning the spirit? If the spirit can be incarnated once, it logically means that it can be reincarnated! It is a conclusion that one necessarily comes to if one follows natural logic, and the only way for these persons to continue to maintain their strange beliefs is to assume that the spirit does not obey the laws of nature; in other words, by installing artificial divisions in their thoughts, which creates confusion.

Again, we just need to follow basic logic to solve what many consider to be great mysteries. All it takes to appreciate the true value of these solutions is a love of simplicity.

IN CONCLUSION

In this book, we have seen how the laws of nature answer the biggest questions. Many other books could be written about how the law clarifies specific topics that we have barely touched, such as the place of humanity in existence, the origin and evolution of the human spirit, the invisible worlds and their inhabitants, the formation of destiny, our responsibility toward existence, and so on. The good news is that such a work already exists. It is the work *In the Light of Truth – The Grail Message*, which was presented in Chapter 11 *(grailmessage.org)*. All these important topics are covered in depth there, and many more. If you want to know more, you can continue your research by reading this work, which will give you infinitely more than I could ever do.

To answer the greatest questions of existence, it is not necessary to rely blindly on religions, or resort to complicated scientific theories, because everything works according to the elementary logic of the laws of nature. These great laws give us all the answers we need, they teach us the meaning of life, as well as the path to follow to integrate ourselves harmoniously into the universal organism.

Bigger is the question, simpler is the answer. Questions about the true nature of consciousness and life are the biggest questions we can ask ourselves, so they must also have the simplest answers. The big answers are in the great laws, the same laws that are at the heart of science. These laws, we all know them intuitively, because we experience them at every moment. To see clearly, all we need is to change our attitude toward the big questions, recognizing that these simple laws, which we already use to guide our everyday life, must also allow us to answer these fundamental questions.

To free our spirit from the mist of artificial confusion, we must seek the light of natural simplicity.

FINAL SUMMARY

*The answers to the great mysteries of science are in the laws of nature.
But to see those answers, we must first leave materialism behind.*

The root of all problems is that we are more concerned with protecting our false beliefs than with seeking logical solutions.

The universality of the laws of nature is the master key.

Humanity only knows the shadow of reality, the essential is invisible.

The essence of reality is energy, not matter.

Consciousness is light perceiving itself.

*Beyond life in visible matter, there is life in invisible matter.
Beyond life in invisible matter, there is life in light.
Beyond life in light, there is Life itself.*

In short: everything is life.

The meaning of life is to serve.

The ideal is to explain a lot with little.

Bigger is the question, simpler is the answer.

The big answers are in the great laws.

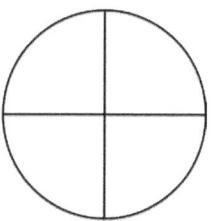

BIBLIOGRAPHY

PHYSICS AND CHEMISTRY

ADAMS, Allan, Matthew EVANS and Barton ZWIEBACH, *Quantum Physics I*, MIT OpenCourseWare, Massachusetts Institute of Technology, spring 2013, < ocw.mit.edu/courses/physics/8-04-quantum-physics-i-spring-2013/ >

BAIS, Sander, *The Equations: Icons of Knowledge*, Harvard University Press, Cambridge, MA, 2005.

—, *Very Special Relativity: An Illustrated Guide,* Harvard University Press, Cambridge, MA, 2007.

BALL, David W., *The Nature of Matter: Understanding the Physical World*, The Teaching Company, Chantilly, VA, < thegreatcourses.com/courses/the-nature-of-matter-understanding-the-physical-world.html >

CARROLL, Sean, *Dark Matter, Dark Energy: The Dark Side of the Universe*, The Teaching Company, Chantilly, VA, < thegreatcourses.com/courses/dark-matter-dark-energy-the-dark-side-of-the-universe.html >

—, *The Higgs Boson and Beyond*, The Teaching Company, Chantilly, VA, < thegreatcourses.com/courses/the-higgs-boson-and-beyond.html >

COHEN-TANNOUDJI, Gilles and Michel SPIRO, *La Matière-Espace-Temps,* Éditions Fayard, Paris, 1986.

CRUISE, Brit, *Journey Into Information Theory*, Khan Academy, < khanacademy.org/computing/computer-science/informationtheory#concept-intro >

DAVIS, Rob B. Jr., *Chemistry and Our Universe: How It All Works*, The Teaching Company, Chantilly, VA, < thegreatcourses.com/courses/chemistry-and-our-universe-how-it-all-works.html >

—, *Foundations of Organic Chemistry*, The Teaching Company, Chantilly, VA, < thegreatcourses.com/courses/foundations-of-organic-chemistry.html >

DUNHAM, William, *Great Thinkers, Great Theorems*, The Teaching Company, Chantilly, VA, <thegreatcourses.com/courses/great-thinkers-great-theorems.html>

DOWNAROWICZ, Tomasz, *Entropy,* Scholarpedia, 2007, <scholarpedia.org/article/Entropy>

DRENNAN, Catherine and Elizabeth VOGEL TAYLOR, *Principles of Chemical Science*, MIT OpenCourseWare, Massachusetts Institute of Technology, fall 2008, <ocw.mit.edu/courses/chemistry/5-111-principles-of-chemical-science-fall-2008/>

FEYNMAN, Richard, *The Very Best of the Feynman Lectures*, Basic Books, New York, 2005.

—, *Quantum Electrodynamics*, Sir Douglas Robb Lectures, University of Auckland, 1979.

—, *The Character of Physical Law*, Messenger Lectures (filmed by the BBC), Cornell University, 1964, <cornell.edu/video/playlist/richard-feynman-messenger-lectures>

FRENCH, Steven, *Identity and Individuality in Quantum Theory*, The Stanford Encyclopedia of Philosophy (Fall 2015 Edition), Edward N. Zalta (ed.), <plato.stanford.edu/archives/fall2015/entries/qt-idind/>

GRAY, Chris G., *Principle of Least Action,* Scholarpedia, 2009, <scholarpedia.org/article/Principle_of_least_action>

GRIBBIN, John, *In Search of Schrödinger's Cat: Quantum Physics and Reality,* Bantam Books, New York, 1984.

GROSSMAN, Jeffrey C., *Thermodynamics: Four Laws That Move the Universe,* The Teaching Company, Chantilly, VA, <thegreatcourses.com/courses/thermodynamics-four-laws-that-move-the-universe.html>

HAWKINS, Stephen, *A Brief History of Time: From the Big Bang to Black Holes*, Bantam Books, New York, 1988.

HOOFT, Gerard 't, *Gauge Theories,* Scholarpedia, 2008, <scholarpedia.org/article/Gauge_theories>

LEWIN, Walter, *Physics II: Electricity and Magnetism*, MIT OpenCourseWare, Massachusetts Institute of Technology, spring 2002.

LINCOLN, Don, *The Theory of Everything: The Quest to Explain All Reality*, The Teaching Company, Chantilly, VA, <thegreatcourses.com/courses/the-theory-of-everything-the-quest-to-explain-all-reality.html>

McBRIDE, J. Michael, *Freshman Organic Chemistry I,* Open Yale Course, Yale University, fall 2008, <oyc.yale.edu/NODE/66>

—, *Freshman Organic Chemistry II,* Open Yale Course, Yale University, spring 2011, <oyc.yale.edu/NODE/71>

MARONEY, Owen, *Information Processing and Thermodynamic Entropy*, The Stanford

Encyclopedia of Philosophy (Fall 2009 Edition), Edward N. Zalta (ed.), <plato.stanford.edu/archives/fall2009/entries/information-entropy/>

MYRVOLD, Wayne, *Philosophical Issues in Quantum Theory,* The Stanford Encyclopedia of Philosophy (Spring 2017 Edition), Edward N. Zalta (ed.), <plato.stanford.edu/archives/spr2017/entries/qt-issues/>

NELSON, Keith A. and Moungi BAWENDI, *Thermodynamic and Kinetics*, MIT OpenCourseWare, Massachusetts Institute of Technology, spring 2008, <ocw.mit.edu/courses/chemistry/5-60-thermodynamics-kinetics-spring-2008/>

PENROSE, Roger, *The Road to Reality: A Complete Guide to the Laws of the Universe,* Jonathan Cape, London, 2004.

RESSLER, Stephen, *Understanding the World's Greatest Structures: Science and Innovation from Antiquity to Modernity*, The Teaching Company, Chantilly, VA, <thegreatcourses.com/courses/understanding-the-world-s-greatest-structures-science-and-innovation-from-antiquity-to-modernity.html>

SCHUMACHER, Benjamin, *Impossible: Physics beyond the Edge*, The Teaching Company, Chantilly, VA, <thegreatcourses.com/courses/impossible-physics-beyond-the-edge.html>

—, *The Science of Information: From Language to Black Holes*, The Teaching Company, Chantilly, VA, <thegreatcourses.com/courses/the-science-of-information-from-language-to-black-holes.html>

SMITH, George, *Newton's Philosophiae Naturalis Principia Mathematica*, The Stanford Encyclopedia of Philosophy (Winter 2008 Edition), Edward N. Zalta (ed.), <plato.stanford.edu/archives/win2008/entries/newton-principia/>

SUSSKIND, Leonard, *The Theoretical minimum: Classical Mechanics*, Stanford Institute for Theoretical Physics, Stanford University, fall 2011, <theoreticalminimum.com/courses/classical-mechanics/2011/fall>

—, *The Theoretical Minimum: Cosmology,* Stanford Institute for Theoretical Physics, Stanford University, winter 2013, <theoreticalminimum.com/courses/cosmology/2013/winter>

—, *The Theoretical Minimum: General Relavity,* Stanford Institute for Theoretical Physics, Stanford University, fall 2012, <theoreticalminimum.com/courses/general-relativity/2012/fall>

—, *The Theoretical Minimum: Particle Physics 1: Basic Concepts,* Stanford Institute for Theoretical Physics, Stanford University, fall 2009, <theoreticalminimum.com/courses/particle-physics-1-basic-concepts/2009/fall>

—, *The Theoretical Minimum: Particle Physics 2: Standard Model,* Stanford Institute for Theoretical Physics, Stanford University, winter 2010, <theoreticalminimum.com/courses/particle-physics-2-standard-model/2010/winter>

—, *The Theoretical Minimum: Particle Physics 3: Supersymmetry and Grand Unification,*

Stanford Institute for Theoretical Physics, Stanford University, spring 2010, <theoreticalminimum.com/courses/particle-physics-3-supersymmetry-and-grand-unification/2010/spring>

—, *The Theoretical Minimum: Quantum Mechanics*, Stanford Institute for Theoretical Physics, Stanford University, winter 2012, <theoreticalminimum.com/courses/quantum-mechanics/2012/winter>

—, *The Theoretical Minimum: Special Relativity and Electrodynamics,* Stanford Institute for Theoretical Physics, Stanford University, spring 2012, <theoreticalminimum.com/courses/special-relativity-and-electrodynamics/2012/spring>

—, *The Theoretical Minimum: Statistical Mechanics,* Stanford Institute for Theoretical Physics, Stanford University, spring 2013, <theoreticalminimum.com/courses/statistical-mechanics/2013/spring>

—, *The Theoretical Minimum: String Theory,* Stanford Institute for Theoretical Physics, Stanford University, fall 2010, <theoreticalminimum.com/courses/string-theory/2010/fall>

SCHWENK, Théodore, *Le chaos sensible (2nd edition),* Éditions Triades, Paris, 1995.

WOLFSON, Richard, *Einstein's Relativity and the Quantum Revolution: Modern Physics for Non-Scientists (2nd edition)*, The Teaching Company, Chantilly, VA, <thegreatcourses.com/courses/einstein-s-relativity-and-the-quantum-revolution-modern-physics-for-non-scientists-2nd-edition.html>

YOUNG, Kenneth, *A Special Lecture: Principle of Least Action*, CUTV, Chinese University of Hong Kong, 2014, <cpr.cuhk.edu.hk/cutv/detail/486?t=prof-kenneth-young-on-a-special-lecture-principle-of-least-action>

ZINN-JUSTIN, Jean and Riccardo GUIDA, *Gauge Invariance,* Scholarpedia, 2008, <scholarpedia.org/article/Gauge_invariance>

BIOLOGY AND ORIGIN OF LIFE

BARD, Jonathan, *Morphogenesis,* Scholarpedia, 2008, <scholarpedia.org/article/Morphogenesis>

BENAROCH, Roy, *Medical School for Everyone: Grand Rounds Cases*, The Teaching Company, Chantilly, VA, <thegreatcourses.com/courses/medical-school-for-everyone-grand-rounds-cases.html>

BEYER, Earl, *Microbiology,* Harriburg Area Community College, 2011–2013, <podcasts.apple.com/us/podcast/biol-221-microbiology-eb/id459057060>

BRANDON, Robert, *Natural Selection,* The Stanford Encyclopedia of Philosophy (Spring 2014 Edition), Edward N. Zalta (ed.), <plato.stanford.edu/archives/spr2014/entries/natural-selection/>

BRIGANDT, Ingo and Alan LOVE, *Reductionism in Biology*, The Stanford Encyclopedia of Philosophy (Spring 2017 Edition), Edward N. Zalta (ed.), <plato.stanford.edu/archives/spr2017/entries/reduction-biology/>

CAIRNS-SMITH, A.G., *Seven Clues to the Origin of Life: A Scientific Detective Story*, Cambridge University Press, Cambridge, UK, 1985.

—, *The origin of life and the nature of the primitive gene*, Journal of Theoretical Biology, vol. 10, no. 1, 1966, pp. 53–88.

CALVIN, Melvin, *Chemical Evolution and the Origin of Life*, University of California, Berkeley, 1955

CONSOLMAGNO, Guy J., *What is Life?*, University of Arizona Science Lectures Series, University of Arizona, 2015, <uascience.org/series/life-in-the-universe/>

DAWKINS, Richard, *The Blind Watchmaker: Why the Evidence of Evolution Reveals a Universe without Design*, W. W. Norton & Company, New York, 1996.

DEGRAZIA, David, *The Definition of Death*, The Stanford Encyclopedia of Philosophy (Spring 2017 Edition), Edward N. Zalta (ed.), <plato.stanford.edu/archives/spr2017/entries/death-definition/>

DEVER Jennifer, *Evolution*, University of San Fransisco, fall 2012, <usfca.edu/catalog/course/414-evolution>

DORNHAUS, Anna R., *Complexity and Evolvability: What Makes Life So Interesting?*, University of Arizona Science Lectures Series, 2015, <uascience.org/series/life-in-the-universe/>

DYSON, Freeman J., *A model for the origin of life*, Journal of Molecular Evolution, vol.18, no. 5, 1982, pp. 344–350.

ENQUIST, Brian J., *Life on Earth: By Chance or By Law?*, University of Arizona Science Lectures Series, University of Arizona, 2015, <uascience.org/series/life-in-the-universe/>

FLEURY, Bruce E., *Mysteries of the Microscopic World*, The Teaching Company, Chantilly, VA, <www.thegreatcourses.com/courses/mysteries-of-the-microscopic-world.html>

GARDEL Claudette, Eric LANDER, Robert WEINBERG and Andrew CHESS, *Introduction to Biology*, MIT OpenCourseWare, Massachusetts Institute of Technology, fall 2004, <ocw.mit.edu/courses/biology/7-012-introduction-to-biology-fall-2004/#>

GODFREY-SMITH, Peter and Kim STERELNY, *Biological Information*, The Stanford Encyclopedia of Philosophy (Summer 2016 Edition), Edward N. Zalta (ed.), <plato.stanford.edu/archives/sum2016/entries/information-biological/>

GRIFFITHS, Paul, *The Distinction Between Innate and Acquired Characteristics*, The Stanford Encyclopedia of Philosophy (Spring 2017 Edition), Edward N. Zalta (ed.), <plato.stanford.edu/archives/spr2017/entries/innate-acquired/>

—, *Philosophy of Biology,* The Stanford Encyclopedia of Philosophy (Spring 2017 Edition), Edward N. Zalta (ed.), <plato.stanford.edu/archives/spr2017/entries/biology-philosophy/>

GUTTINGER, Stephan and John DUPRÉ, *Genomics and Postgenomics,* The Stanford Encyclopedia of Philosophy (Winter 2016 Edition), Edward N. Zalta (ed.), <plato.stanford.edu/archives/win2016/entries/genomics/>

HAKEN, Hermann, *Self-organization,* Scholarpedia, 2008, <scholarpedia.org/article/Self-organization>

HAROLD, Franklin M., *In Search of Cell History: The Evolution of Life's Building Blocks,* The University of Chicago Press, University of Chicago, 2014.

HAZEN, Robert M., *Origins of Life,* The Teaching Company, Chantilly, VA, <thegreatcourses.com/courses/origins-of-life.html>

HORDIJK, Wim, Jotun HEIN and Mike STEEL, *Autocatalytic Sets and the Origin of Life,* Entropy, vol. 12, no. 7, 2010, pp. 1733–1742.

IZHIKEVICH, Eugene M., John H. Conway and Anil Seth, *Game of Life,* Scholarpedia, 2015, <scholarpedia.org/article/Game_of_Life>

KAPLAN, Ken, *The Inner Workings of Cells,* University of California, Davis, winter 2008.

KELSO, J. A. Scott, *Synergies,* Scholarpedia, 2008, <scholarpedia.org/article/Synergies>

LANE, Nick, John F. ALLEN and William MARTIN, *How did LUCA make a living? Chemiosmosis in the origin of life,* BioEssays, vol. 32, no. 4, 2010, pp. 271–280.

LANE, Nick, *The Vital Question: Energy, Evolution and the Origin of Complex Life,* W. W. Norton & Company, New York, 2015.

LAURETTA, Dante S., *Planet Formation and the Origin of Life,* University of Arizona Science Lectures Series, University of Arizona, 2015, <uascience.org/series/life-in-the-universe/>

LENNOX, James, *Darwinism,* The Stanford Encyclopedia of Philosophy (Spring 2017 Edition), Edward N. Zalta (ed.), <plato.stanford.edu/archives/spr2017/entries/darwinism/>

LEVY, Matthew and Stanley L. MILLER, *The stability of the RNA bases: Implications for the origin of life,* Proceedings of the National Academy of Sciences, vol. 95, no. 14, 1998, pp. 7933–7938

LUNINE, Jonathan, *Life's Extreme Edge: The Limits of Organic Life on Earth and Other Planets,* University of Arizona Science Lectures Series, University of Arizona, 2008, <uascience.org/series/the-edges-of-life/>

MAURETTE, M., *Carbonaceous Micrometeorites and the Origin of Life,* Origins of Life and Evolution of the Biosphere, vol. 28, no. 4–6, 1998, pp. 385–412.

McFARLAND, Ben, *Biochemistry 4361,* Seattle Pacific University, fall 2015, <podcasts.apple.com/pk/itunes-u/biochemistry-bio-chem-4361-fall-2015/id1044873866?mt=10>

—, *Biochemistry 4362,* Seattle Pacific University, winter 2016, <podcasts.apple.com/kw/podcast/biochemistry-bio-chem-4362-winter-2016/id1073253861>

MARTIN, William, John BAROSS, Deborah KELLEY and Michael J. RUSSELL, *Hydrothermal vents and the origin of life*, Nature Reviews Microbiology, vol. 6, no. 11, 2008, pp. 805–814.

MAZUR, Suzan, *The Origin of Life Circus: A How To Make Life Extravaganza*, Caswell Books, New York, 2014.

MESLER, Bill and H. James CLEAVES II, *A Brief History of Creation: Science and the Search for the Origin of Life,* W. W. Norton & Company, New York, 2016.

MILLER, Stanley L. and Jeffrey L. BADA, *Submarine hot springs and the origin of life*, Nature, vol. 334, no. 6183, 1988, pp. 609–611.

MONACO, Paul, *Genetics*, Quillen College of Medecine, East Tennesse State University, 2015,<podcasts.apple.com/us/podcast/genetics/id384931324>

MONOD, Jacques, *Le hasard and la nécessité: Essai sur la philosophie naturelle de la biologie moderne*, Éditions du Seuils, Paris, 1970.

NITSCHKE, W. and M.J. RUSSELL, *Just Like the Universe the Emergence of Life had High Enthalpy and Low Entropy Beginnings*, Journal of Cosmology, vol. 10, 2010, pp. 3200–3216.

ORGEL, Leslie E., *The origin of life: a review of facts and speculations*, Trends in Biochemical Sciences, vol. 23, no. 12, 1998, pp. 491–495.

PACE, Norman R. and Terry L. MARSH, *Rna catalysis and the origin of life*, Origins of Life and Evolution of the Biosphere, vol. 16, no. 2, 1985, pp. 97–116.

PRICE, Pp. Buford, *Microbial life in glacial ice and implications for a cold origin of life*, FEMS Microbiology Ecology, vol. 59, no. 2, 2007, pp. 217–231.

PROSS, Addy, *Causation and the Origin of Life: Metabolism or Replication First?*, Origins of Life and Evolution of the Biosphere, vol. 34, no. 3, 2004, pp. 307–321.

—, *What is Life?: How Chemistry Become Biology,* Oxford University Press, Oxford, 2012.

RUSSELL, Michael J., *First Life*, American Scientist, vol. 94, 2006, pp. 32–39.

RUSSELL, Michael J. and Allan J. HALL, *From Geochemistry to Biochemistry: Chemiosmotic coupling and transition element clusters in the onset of life and photosynthesis*, The Geochemical News, no. 113, 2002, pp. 6–12.

SAGAN, Carl, *On the Origin and Planetary Distribution of Life*, Radiation Research, vol. 15, no. 2, 1961, pp. 174–192.

SARKAR, Sahotra, *Ecology*, The Stanford Encyclopedia of Philosophy (Winter 2016 Edition), Edward N. Zalta (ed.), <plato.stanford.edu/archives/win2016/entries/ecology/>

SCHRODINGER, Erwin, *What Is Life?: The Physical Aspect of the Living Cell,* Cambridge University Press, Cambridge, UK, 1944.

SHAPIRO, James A., *Evolution: A View from the 21st Century*, FT Press Science, Upper Saddle River, New Jersey, 2011.

SHAPIRO, Robert, *A Replicator Was Not Involved in the Origin of Life*, IUBMB Life, vol. 49, no. 3, 2000. pp. 173–176.

—, *Origins: A Skeptic's Guide to the Creation of Life on Earth,* , Bantam Books, New York, 1987.

—, *Small Molecule Interactions were Central to the Origin of Life*, The Quarterly Review of Biology, vol. 81, no. 2, 2006, pp. 105–126.

SZATHMARY, Eörs and László DEMETER, *Group selection of early replicators and the origin of life*, Journal of Theoretical Biology, vol.128, no.4, 1987, pp. 463–486.

TABERY, James, Monika PIOTROWSKA and Lindley DARDEN, *Molecular Biology*, The Stanford Encyclopedia of Philosophy (Spring 2017 Edition), Edward N. Zalta (ed.), <plato.stanford.edu/archives/spr2017/entries/molecular-biology/>

TAYLOR, Peter and Richard LEWONTIN, *The Genotype/Phenotype Distinction*, The Stanford Encyclopedia of Philosophy (Summer 2017 Edition), Edward N. Zalta (ed.), <plato.stanford.edu/archives/sum2017/entries/genotype-phenotype/>

UREY, Harold C., *On the Early Chemical History of the Earth and the Origin of Life*, Proceedings of the National Academy of Sciences, vol. 38, no. 4, 1952, pp. 351–363.

VASAS, Vera, Eörs SZATHMARY and Mauro SANTOS, *Lack of evolvability in self-sustaining autocatalytic networks constraints metabolism-first scenarios for the origin of life*, Proceedings of the National Academy of Sciences, vol. 107, no. 4, 2010, pp. 1470–1475.

WACHTERSHAUSER, Günter, *The Origin of Life and its Methodological Challenge*, Journal of Theoretical Biology, vol. 187, no. 4, 1997, pp. 483–494.

WATERS, Ken, *Molecular Genetics*, The Stanford Encyclopedia of Philosophy (Fall 2013 Edition), Edward N. Zalta (ed.), <plato.stanford.edu/archives/fall2013/entries/molecular-genetics/>

WEBER, Bruce, *Life*, The Stanford Encyclopedia of Philosophy (Spring 2015 Edition), Edward N. Zalta (ed.), <plato.stanford.edu/archives/spr2015/entries/life/>

WEBER, Marcel, *Experiment in Biology*, The Stanford Encyclopedia of Philosophy (Winter 2014 Edition), Edward N. Zalta (ed.), <plato.stanford.edu/archives/win2014/entries/biology-experiment/>

WILKINS, John S. and David HULL, *Replication and Reproduction*, The Stanford Encyclopedia of Philosophy (Spring 2014 Edition), Edward N. Zalta (ed.), <plato.stanford.edu/archives/spr2014/entries/replication/>

YOCKEY, Hubert Pp., *Origin of life on earth and Shannon's theory of communication*, Computers & Chemistry, vol. 24, no. 1, 2000, pp. 105–123.

—, *Self organization origin of life scenarios and information theory*, Journal of Theoretical Biology, vol. 91, no. 1, 1981, pp. 13–31.

NEUROSCIENCE AND CONSCIOUSNESS

AGUIRRE, Geoffrey, *Brain Imaging, Reality and Hype*, Centre for Cognitive Neuroscience, University of Pennsylvania, 2010, <podcasts.apple.com/us/podcast/center-for-cognitive-neuroscience/id432575137>

ATMANSPACHER, Harald, *Quantum Approaches to Consciousness*, The Stanford Encyclopedia of Philosophy (Summer 2015 Edition), Edward N. Zalta (ed.), <plato.stanford.edu/archives/sum2015/entries/qt-consciousness/>

BAARS, Bernard J., *Global workspace theory of consciousness: toward a cognitive neuroscience of human experience*, The Boundaries of Consciousness: Neurobiology and Neuropathology, vol. 150, pp. 45–53.

BARTTFELD, Pablo, Lynn UHRIG, Jacobo D. SITT, Mariano SIGMAN, Béchir JARRAYA and Stanislas DEHAENE, *Signature of consciousness in the dynamics of resting-state brain activity*, Proceedings of the National Academy of Sciences, vol. 112, no. 3, 2015, pp. 887–892.

BEAUMONT, Graham J., *Introduction to Neuropsychology (2nd edition)*, The Guilford Press, New York, 2008.

BERZHANSKAYA, Julia and Giorgio ASCOLI, *Computational Neuroanatomy,* Scholarpedia, 2008, <scholarpedia.org/article/Computational_neuroanatomy>

BEYER, Earl, *Neurobiology*, Harrisburg Area Community College, winter 2014, <podcasts.apple.com/us/podcast/neurobiology/id780663908>

BISIACH, Edoardo, Erminio CAPITANI, Claudio LUZZATTI and Daniela PERANI, *Brain and conscious representation of outside reality*, Neuropsychologia, vol. 19, no. 4, 1981, pp. 543–551.

BLACKMORE, Susan, *Conversations on Consciousness: What the Best Minds Think about the Brain, Free Will, and What It Means to Be Human,* Oxford University Press, Oxford, 2006

—, *Near-Death Experiences*, Journal of the Royal Society of Medicine, vol. 89, no. 2, 1996, pp. 73–76.

—, *State of the Art – The Psychology of Consciousness,* The psychologist, vol. 14, 2001, pp. 522–525.

BLANKE, Olaf, *Multisensory brain mechanisms of bodily self-consciousness*, Nature Reviews Neuroscience, vol. 13, no. 8, 2012, pp. 556–571.

BICKLE, John, Peter MANDIK and Anthony LANDRETH, *The Philosophy of Neuroscience*, The Stanford Encyclopedia of Philosophy (Summer 2012 Edition), Edward N. Zalta (ed.), <plato.stanford.edu/archives/sum2012/entries/neuroscience/>

BLOOM, Pail, *Introduction to Psychology,* Open Yale Course, Yale University, spring 2007, <oyc.yale.edu/NODE/231>

BOLY, Melanie, and al., *Intrinsic Brain Activity in Altered States of Consciousness: How Conscious*

Is the Default Mode of Brain Function?, Annals of the New York Academy of Sciences, vol. 1129, no. 1, 2008, pp. 119–129.

BRAITENBERG, Valentino, *Brain,* Scholarpedia, 2007, < scholarpedia.org/article/Brain >

BROOK, Andrew and Paul RAYMONT, *The Unity of Consciousness*, The Stanford Encyclopedia of Philosophy (Summer 2017 Edition), Edward N. Zalta (ed.), < plato.stanford.edu/archives/sum2017/entries/consciousness-unity/ >

BURKE, Robert E., *Spinal Cord,* Scholarpedia, 2008, < scholarpedia.org/article/Spinal_cord >

BURKEMAN, Oliver, *Why can't the world's greatest minds solve the mystery of consciousness ?*, < theguardian.com/science/2015/jan/21/-sp-why-cant-worlds-greatest-minds-solve-mystery-consciousness >, 21 janvier 2015.

BYRNE, Alex and Sinha PAWAN, *Philosophical Issues in Brain Science*, MIT OpenCourseWare, Massachusetts Institute of Technology, spring 2009, < ocw.mit.edu/courses/linguistics-and-philosophy/24-08j-philosophical-issues-in-brain-science-spring-2009/# >

CHALMERS, David J., *The Conscious Mind : In Search of a Fundamental Theory,* Oxford University Press, Oxford, 1996.

—, *Philosophy of Mind : Classical and Contemporary Readings,* Oxford University Press, Oxford, 2002.

DEHAENE, Stanislas, *Towards a cognitive neuroscience of consciousness : basic evidence and a workspace framework*, Cognition, vol. 79, no. 1–2, 2001, pp. 1–37.

DENNET, Daniel C., *Consciousness Explained,* Little, Brown & Company, New York, 1991.

DORNHAUS Anna, *Evolution of Mind and Brain*, University of Arizona Science Lectures Series, University of Arizona, 2010, < uascience.org/series/mind-and-brain/ >

GERKEN, LouAnn, *The Making of a Mind*, University of Arizona Science Lectures Series, University of Arizona, 2010, < uascience.org/series/mind-and-brain/ >

GIACINO, Joseph T., Joseph J. FINS, Steven LAUREYS and Nicholas SCHIFF, *Disorders of consciousness after acquired brain injury : the state of the science*, Nature Reviews Neurology, vol. 10, no. 2, 2014, pp. 99–114.

GOFF, Philip, William SEAGER and Sean ALLEN-HERMANSON, *Panpsychism*, The Stanford Encyclopedia of Philosophy (Fall 2017 Edition), Edward N. Zalta (ed.), < plato.stanford.edu/archives/fall2017/entries/panpsychism/ >

GREYSON, Bruce, *The Near-Death Experience Scale : Construction, Reliability, and Validity*, The Journal of Nervous and Mental Disease, vol. 171, no. 6, 1983, pp. 369–375.

GRIM, Patrick, *Mind-Body Philosophy*, The Teaching Company, Chantilly, VA, < thegreatcourses.com/courses/mind-body-philosophy.html >

—, *Philosophy of Mind : Brains, Consciousness, and Thinking Machines*, The Teaching

Company, Chantilly, VA, <thegreatcourses.com/courses/philosophy-of-mind-brains-consciousness-and-thinking-machines.html>

GROSSBERG, Stephen, *The Link between Brain Learning, Attention, and Consciousness*, Consciousness and Cognition, vol. 8, no.1, 1999, pp. 1–44.

HAMEROFF, Stuart, *Quantum computation in brain microtubules? The Penrose–Hameroff 'Orch OR' model of consciousness*, Philosophical Transactions of the Royal Society A, vol. 356, 1998, pp. 1869–1896.

JO, Han-Gue, Marc WITTMANN, Tilmann L. BORGHARDT, Thilo HINTERBERGER and Stefan SCHMIDT, *First-person approaches in neuroscience of consciousness: Brain dynamics correlate with the intention to act*, Consciousness and Cognition, vol. 26, 2014, pp. 105–116.

KASZNIAK, Alfred W., *Metamemory: How Does the Brain Predict Itself?*, University of Arizona Science Lectures Series, University of Arizona, 2010, <uascience.org/series/mind-and-brain/>

KOCH, Christof and Naotsugu TSUCHIYA, *Attention and consciousness: two distinct brain processes*, Trends in Cognitive Sciences, vol. 11, no. 1, 2007, pp. 16–22.

KURZWEIL, Ray, *How to Create a Mind: The Secret of Human Thought Revealed*, Penguin Books, New York, 2012.

LAUREYS, Steven, Adrian M. OWEN and Nicholas D. SCHIFF, *Brain function in coma, vegetative state, and related disorders*, The Lancet Neurology, vol. 3, no. 9, 2004, pp. 537–546.

LEVIN, Janet, *Functionalism*, The Stanford Encyclopedia of Philosophy (Winter 2017 Edition), Edward N. Zalta (ed.), <plato.stanford.edu/archives/win2017/entries/functionalism/>

LLINAS, Rodolfo, *Neuron,* Scholarpedia, 2008, <scholarpedia.org/article/Neuron>

LLINAS, Rodolfo and Mario N. NEGRELLO, *Cerebellum,* Scholarpedia, 2015, <scholarpedia.org/article/Cerebellum>

LLINAS, Rodolfo and Urs RIBARY, *Consciousness and the Brain: The Thalamocortical Dialogue in Health and Disease*, Annals of the New York Academy of Sciences, vol. 929, no. 1, 2006, pp.166–175.

MICHAUD, Yves, *La matière et la conscience,* Le Québec sceptique, no. 98, 2019, pp. 13–19.

MOBBS, Dean and Caroline WATT, *There is nothing paranormal about near-death experiences: how neuroscience can explain seeing bright lights, meeting the dead, or being convinced you are one of them*, Trends in Cognitive Sciences, vol. 15, no. 10, 2011, pp. 447–449.

MOODY, Raymond, *La vie après la vie: Enquête à propos d'un phénomène: la survie de la conscience après la mort du corps*, Éditions Robert Laffont, Paris, 1977.

MORSE, Stephen, *Free Will, Responsibility and Brain Function*, Centre for Cognitive Neuroscience, University of Pennsylvania, 2010, <podcasts.apple.com/us/podcast/center-for-cognitive-neuroscience/id432575137>

NADEL, Lynn, *Building Brains, Making Minds,* University of Arizona Science Lectures Series, University of Arizona, 2010, < uascience.org/series/mind-and-brain/ >

NAGEL, Thomas, *Brain bisection and the unity of consciousness*, Synthese, vol. 22, no. 3–4, 1971, pp. 396–413.

NICHOLS, Shaun, *Morality and the Emotional Brain*, University of Arizona Science Lectures Series, University of Arizona, 2010, < uascience.org/series/mind-and-brain/ >

PARNIA, Sam and Peter FENWICK, *Near death experiences in cardiac arrest: visions of a dying brain or visions of a new science of consciousness*, Resuscitation, vol. 52, no. 1, 2002, pp. 5–11.

PENFIELD, Wilder, *The Interpretive Cortex: The stream of consciousness in the human brain can be electrically reactivated*, Science, vol. 129, no. 3365, 1959, pp. 1719–1725.

PENROSE, Roger, *The Emperor's New Mind: Concerning Computers, Minds, and the Laws of Physics,* Oxford University Press, Oxford, 1989.

RAMSEY, William, *Eliminative Materialism*, The Stanford Encyclopedia of Philosophy (Winter 2016 Edition), Edward N. Zalta (ed.), < plato.stanford.edu/archives/win2016/entries/materialism-eliminative/ >

REDGRAVE, Peter, *Basal ganglia,* Scholarpedia, 2007, < scholarpedia.org/article/Basal_ganglia >

ROBINSON, Daniel N., *Consciousness and Its Implications*, The Teaching Company, Chantilly, VA, < thegreatcourses.com/courses/consciousness-and-its-implications.html >

ROBINSON, Howard, *Dualism*, The Stanford Encyclopedia of Philosophy (Fall 2017 Edition), Edward N. Zalta (ed.), < plato.stanford.edu/archives/fall2017/entries/dualism/ >

ROBINSON, William, *Epiphenomenalism*, The Stanford Encyclopedia of Philosophy (Fall 2015 Edition), Edward N. Zalta (ed.), < plato.stanford.edu/archives/fall2015/entries/epiphenomenalism/ >

SACKS, Oliver, *The Man Who Mistook His Wife For A Hat: And Other Clinical Tales*, Touchstone, New York, 1998.

SCHNEIDER, Gerald E., *Neuroscience and Behavior*, MIT OpenCourseWare, Massachusetts Institute of Technology, fall 2003, < ocw.mit.edu/courses/brain-and-cognitive-sciences/9-01-neuroscience-and-behavior-fall-2003/ >

SCHLOSSER, Markus, *Agency*, The Stanford Encyclopedia of Philosophy (Winter 2019 Edition), Edward N. Zalta (ed.), < plato.stanford.edu/archives/win2019/entries/agency/ >

SCHUZ, Almut, *Neuroanatomy,* Scholarpedia, 2008, < scholarpedia.org/article/Neuroanatomy >

SETH, Anil, *Models of Consciousness,* Scholarpedia, 2007, < scholarpedia.org/article/Models_of_consciousness >

SHERMAN, S. Murray, *Thalamus,* Scholarpedia, 2006, < scholarpedia.org/article/Thalamus >

SKIDMORE, Frank, *Imaging the Brain*, University of Warwick, 2010, <podcasts.apple.com/us/podcast/imaging-the-brain/id491053842>

SMITH, David Woodruff, *Phenomenology*, The Stanford Encyclopedia of Philosophy (Winter 2016 Edition), Edward N. Zalta (ed.), <plato.stanford.edu/archives/win2016/entries/phenomenology/>

STUFFLEBEAM, Robert, *Philosophy of Mind*, University of New Orleans, 2013, <podcasts.apple.com/course/philosophy-of-mind/id662140305>

SCHWARTZ, Jeffrey M., Henry P. STAPP and Mario BEAUREGARD, *Quantum physics in neuroscience and psychology: a neurophysical model of mind-brain interaction*, Philosophical Transactions of the Royal Society B: Biological Sciences, vol. 360, no. 1458, 2005, pp. 1309–1327.

TOLBERT, Leslie P., *The Plastic Brain*, University of Arizona Science Lectures Series, University of Arizona, 2010, <uascience.org/series/mind-and-brain/>

TONONI, Guilio, *Consciousness, information integration, and the brain*, Progress in Brain Research, vol. 150, 2005, pp. 109–126.

TYE, Michael, *Qualia*, The Stanford Encyclopedia of Philosophy (Winter 2016 Edition), Edward N. Zalta (ed.), <plato.stanford.edu/archives/win2016/entries/qualia/>

VAN GULICK, Robert, *Consciousness*, The Stanford Encyclopedia of Philosophy (Spring 2014 Edition), Edward N. Zalta (ed.), <plato.stanford.edu/archives/spr2014/entries/consciousness/>

VELMANS, Max, *Consciousness and the "causal paradox"*, Behavioral and Brain Sciences, vol. 19, special no. 3, 1996, pp. 538–542.

—, *Consciousness, brain and the physical world*, Philosophical Psychology, vol. 3, no. 1, 1990, pp. 77–99.

WU, Wayne, *The Neuroscience of Consciousness*, The Stanford Encyclopedia of Philosophy (Winter 2018 Edition), Edward N. Zalta (ed.), <plato.stanford.edu/archives/win2018/entries/consciousness-neuroscience/>

SCIENCE AND PHILOSOPHY

ANDERSEN, Hanne and Brian HEPBURN, *Scientific Method*, The Stanford Encyclopedia of Philosophy (Summer 2016 Edition), Edward N. Zalta (ed.), <plato.stanford.edu/archives/sum2016/entries/scientific-method/>

ANGIER, Natalie, *The Canon: A Whirligig Tour of the Beautiful Basics of Science*, Houghton Mifflin Company, New York, 2007.

ATCHISON, Jarrod, *The Art of Debate*, The Teaching Company, Chantilly, VA, <thegreatcourses.com/courses/the-art-of-debate.html>

BAILLARGEON, Normand, *Petit cours d'autodéfense intellectuelle,* Lux Éditeur, Montréal, 2006.

BENJAMIN, Craig G., *Big History of Civilizations*, The Teaching Company, Chantilly, VA, <thegreatcourses.com/courses/the-big-history-of-civilizations.html>

BLACKBURN, Pierre, *Logique de l'argumentation (2nd edition)*, Éditions du Renouveau Pédagogique, Saint-Laurent,1994.

BRYSON, Bill, *A Short History of Nearly Everything*, Broadway Books, New York, 2003.

CARROLL, John W., *Laws of Nature*, The Stanford Encyclopedia of Philosophy (Fall 2016 Edition), Edward N. Zalta (ed.), <plato.stanford.edu/archives/fall2016/entries/laws-of-nature/>

CHRISTIAN, David, *Big History: The Big Bang, Life on Earth, and the Rise of Humanity*, The Teaching Company, Chantilly, VA, <thegreatcourses.com/courses/big-history-the-big-bang-life-on-earth-and-the-rise-of-humanity.html>

DAWKINS, Richard, *The God Delusion,* Mariner Books, New York, 2008.

DE CRUZ, Helen, *Religion and Science*, The Stanford Encyclopedia of Philosophy (Spring 2017 Edition), Edward N. Zalta (ed.), <plato.stanford.edu/archives/spr2017/entries/religion-science/>

DE ROSNAY, Joël, *Le macroscope: vers une vision globale*, Éditions du Seuil, Paris, 1975.

DOWNES, Stephen, *Stephen's Guide to the Logical Fallacies*, 1995-2010, <fallacies.ca/toc.htm>

FLEW, Anthony, *There is a God: How the world's most notorious atheist changed his mind*, HarperOne, New York, 2008.

GOLDMAN, Steven L., *Great Scientific Ideas That Changed the World*, The Teaching Company, Chantilly, VA, <thegreatcourses.com/courses/great-scientific-ideas-that-changed-the-world.html>

—, *Science Wars: What Scientists Know and How They Know It*, The Teaching Company, Chantilly, VA, <thegreatcourses.com/courses/science-wars-what-scientists-know-and-how-they-know-it.html>

GREGORY, Frederick, *History of Science: 1700-1900,* The Teaching Company, Chantilly, VA, <thegreatcourses.com/courses/history-of-science-1700-1900.html>

GRUSH, Rick, *Introduction to Logic*, University of California, San Diego, fall 2006.

HANSEN, Hans, *Fallacies*, The Stanford Encyclopedia of Philosophy (Fall 2017 Edition), Edward N. Zalta (ed.), <plato.stanford.edu/archives/fall2017/entries/fallacies/>

HITCHENS, Christopher, *God Is Not Great: How Religion Poisons Everything*, McClelland & Stewart, Toronto, 2007.

HOLMES, Arthur, *A History of Philosophy*, Wheaton College, 2014, <podcasts.apple.com/podcast/a-history-of-philosophy-by-dr-arthur-f-holmes/id903437872>

KUHN, Thomas S., *The Structure of Scientific Revolutions*, University of Chicago Press, Chicago, 1996.

LORANGER, Jessy and Jonathan, *L'évaluation de la scientificité et le mythe des pseudosciences : La lumière sur les enjeux non-scientifiques régissant la crédibilité scientifique et sur le piège des faux sceptiques*, Éditions Propulsion, Repentigny, QC, 2018.

MARTIN, Dale B., *Introduction to New Testament history and literature*, Open Yale Courses, 2009, < podcasts.apple.com/ca/podcast/introduction-to-new-testament-history-literature-video/id341652026 >

MURRAY, Elizabeth A., *Trails of Evidence : How Forensic Science Works*, The Teaching Company, Chantilly, VA, < thegreatcourses.com/courses/trails-of-evidence-how-forensic-science-works.html >

NASH, Ronald H., *History of Philosophy and Christian Tought*, Reformed Theological Seminary, 2010, < podcasts.apple.com/ca/podcast/history-of-philosophy-and-christian-thought/id403537295 >

—, *Modern Philosophy*, Reformed Theological Seminary, 2011, < podcasts.apple.com/ca/podcast/modern-philosophy/id412910038 >

NOVELLA, Steven, *Your Deceptive Mind : A Scientific Guide to Critical Thinking Skills*, The Teaching Company, Chantilly, VA, < thegreatcourses.com/courses/your-deceptive-mind-a-scientific-guide-to-critical-thinking-skills.html >

POPPER, Karl, *Conjectures and Refutations : The Growth of Scientific Knowledge,* Basic Books, New York, 1962.

PRINCIPE, Lawrence M., *History of Science : Antiquity to 1700*, The Teaching Company, Chantilly, VA, < thegreatcourses.com/courses/history-of-science-antiquity-to-1700.html >

RAIA, Courtenay, *History 2D : Science, Magic and religion*, University of California, Los Angeles, spring 2009.

SAGAN, Carl, *The Demon-Haunted World : Science as a Candle in the Dark*, Ballantine Books, New York, 1996.

SCHAEFER, Bradley E., *The Remarkable Science of Ancient Astronomy*, The Teaching Company, Chantilly, VA, < thegreatcourses.com/courses/the-remarkable-science-of-ancient-astronomy.html >

STUFFLEBEAM, Robert, *Introduction to Logic*, University of New Orleans, 2015, < podcasts.apple.com/course/introduction-to-logic/id1074056829 >

—, *Philosophy of the Natural Sciences*, University of New Orleans, 2014, < podcasts.apple.com/us/course/philosophy-of-the-natural-sciences/id901197271 >

TALBOT, Marianne, *Critical Reasoning for Beginners*, Oxford university, 2010, < podcasts.ox.ac.uk/series/critical-reasoning-beginners >

TREFIL, James, *The Nature of Science: An A-Z Guide to the Laws & Principles Governing Our Universe,* Houghton Mifflin Company, New York, 2003

VOTH, Grant L., Kathryn McCLYMOND, Julius H. BAILEY and Robert Andre LAFLEUR, *Great Mythologies of the World,* The Teaching Company, Chantilly, VA, <thegreatcourses.com/courses/great-mythologies-of-the-world.html>

WAINWRIGHT, William, *Concepts of God,* The Stanford Encyclopedia of Philosophy (Spring 2017 Edition), Edward N. Zalta (ed.), <plato.stanford.edu/archives/spr2017/entries/concepts-god/>

WARBURTON, Nigel, *Philosophy: The Classics,* Wizzard Media, 2007 <podcasts.apple.com/podcast/philosophy-the-classics/id254465298>

WESTON, Anthony, *A Rulebook for Arguments (3rd edition),* Hackett Publishing Company, Indianapolis, 2000.

WINTHER, Rasmus Grønfeldt, *The Structure of Scientific Theories,* The Stanford Encyclopedia of Philosophy (Winter 2016 Edition), Edward N. Zalta (ed.), <plato.stanford.edu/archives/win2016/entries/structure-scientific-theories/>

WOODWARD, James, *Scientific Explanation,* The Stanford Encyclopedia of Philosophy (Fall 2017 Edition), Edward N. Zalta (ed.), <plato.stanford.edu/archives/fall2017/entries/scientific-explanation/>

www.ingramcontent.com/pod-product-compliance
Lightning Source LLC
Chambersburg PA
CBHW020635220526
45464CB00001B/164